An Ice A

Susan Thomas lives in Oxford. She graduated with an Honours degree in psychology with statistics, and has a Masters degree in criminology. Hobbies include reading cosmology and astrophysics and divining with the ancient trigrams of the I Ching.

Contents:

Introduction

The Earth is now on the brink of entering a cataclysmic Ice Age. Compelling evidence indicates that the warm, twelve thousand year-long Holocene interglacial will soon be coming to an end. The Earth will soon return to ice age conditions. This will be a major ice age and not the mini ice age that has been widely discussed in the media. During the last major ice age much of the land mass of Great Britain was buried beneath ice. The ice sheet was a mile thick in places!

Every 100, 000 years our Earth enters a new ice age covering vast tracts of the northern land masses including America and Great Britain. The Northern Hemisphere due to its large land mass is inherently more vulnerable to an ice age as snow settles on the ground. The warning signs that are pointing to a new glaciation event in the Northern Hemisphere will be explored in this book. The mother of all ice ages may arrive by the year 2050.

There has been much publicity given recently to the return of the mini ice age conditions last seen in the seventeenth century. I have long surmised that the mini ice age will be the tipping point. The mini ice age will facilitate the arrival of our major ice age which is due soon according to irrefutable geologic records. It is likely that that the great ice age will commence at the end of the mini ice age in the year 2050.The next ice age is now due and could arrive within decades.

I have been researching this cold bleak scenario for twenty years and now feel confident, and also troubled, that the ice age is about to arrive. Think of the chaos that erupted in 2009 when the city of London ground to a halt due to unexpected snowfalls in February. The hapless Mayor of London, Boris Johnson, was tearing his blonde hair out as trains and buses ceased to operate. Commuters were left stranded and unable to reach their offices. Now multiply this chaos by one hundred and you can imagine how an ice age might unfold!

It is well known that statisticians can play with numbers to assemble graphs of climate. I am something of a statistician myself! However the geologic record of our ice ages is set in stone and ice and cannot be disputed. The temperature graphs derived from geologic records depict the massive time scales of previous ice ages. The irrefutable records show that we are overdue the expected ice age.

The million year long geological records indicate that our major ice age is coming, and no amount of trace greenhouse gases will be able to prevent it. Interestingly these paleoclimatic graphs also show a sudden rise in temperatures just before the ice age commences. Might our present warm spells of weather and the Arctic amplification events be another warning sign of an impending ice age?

These days a lot of people are genuinely concerned about the environment and climate change. With the focus on green house gases the glaring fact that we are now expecting a major ice age has somehow been overlooked. While the media attention has been diverted to an anthropogenic global warming, the huge elephant in the room was being completely ignored. This white elephant is our overdue ice age!

Every winter there are in excess of 40, 000 extra winter deaths in the United Kingdom related to cold weather. Yet the recent focus has been entirely on a runaway global warming. As far as I am aware there have been far fewer deaths linked to warm weather in our country. I now believe that there are quite a few warning indicators of a major ice age arriving very soon and so decided to write this book immediately.

Here in Great Britain we have far more to fear from the unexpected arrival of our overdue ice age! During the last ice age the ice sheet miraculously stopped just north of Oxford. This fact is rather serendipitous for an academic city where the concept of global warming has taken off in a big way! Oxford even has its own magazine the "Anthroposphere" devoted to the concept that we are warming the climate. While the southern part of our small country may possibly breathe a sigh of relief things look very severe in the north. If the pattern of the last ice age is repeated populated thriving cities such as Glasgow and Manchester and the whole of Wales will be buried beneath tons of ice!

Indeed I recall meeting with Sir David King, former Chief Scientific Adviser to the British Government, at a climate change event in Oxford. He was at the inauguration of Oxford`s new college, the Martin 21st Century School. We had a little joke together about the possibility of another ice age and he said it would not be much fun for the football supporters of Old Trafford if an ice age came, since Manchester city was buried under a mile of ice during the last ice age!

Due to factors to be discussed in this thoroughly researched book, there is the distinct possibility that our expected ice age will be even more deadly than the last ice age. Great Britain is uniquely vulnerable to a drop in temperatures due to its northern latitude. It is easy to forget that we sited on the same latitude as snowy cities in Canada. The mild climate of Great Britain benefits from the warmth of the Gulf Stream as well as the prevailing westerly maritime winds.

This book will explore the very real possibility of a new ice age in our lifetime as these benign climatic influences are disrupted and cold north east winds prevail. Think of the huge difference the change to a cold east wind direction made during the snowy event of early March 2018, when the Siberian "Beast from the East" roared in. These cold wind patterns are predicted to increase as the solar sunspot cycle powers down.

Most of America, Scandinavia, Northern Europe and the British Isles will eventually become buried under thick ice sheets American survivalists also known as preppers will be in a good position to weather the icy storms with their self sufficient life style and acres of rural land. Here in our small island we will need plenty of snow clearing technology to keep Great Britain working. Unless we act now there will be lethal blackouts as wind and solar technology are unable to function in freezing temperatures. By the middle of the century all hopes of a grape growing climate will be long forgotten. It will become the era of ice age Britain!

Chapter 1: Ice Age Earth and the Milankovitch Cycles

A NASA satellite in Antarctica recorded the coldest temperature on our planet, an astonishing low of minus 94.7 C. There has been so much publicity about anthropogenic global warming that many readers will be surprised to learn that we are technically still living in an ice age at the present time! This present ongoing great ice age is called the Quarternary ice age.

This ice age started 2.58 million years ago. Surprisingly this Quarternary ice age is the coldest period on Earth for the last 500 million years! These long time scales help imbue a sense of perspective that is often lacking when discussions of Earth`s climate take place. The entire Quaternary Period is referred to as an ice age because at least one permanent large ice sheet, such as the Antarctic ice sheet, has existed continuously during this time. The Arctic ice became a fixture later, at around two and a half million years ago. For the last two and half million years there has been ice at both poles and plentiful snow accumulations elsewhere.

Throughout the Quarternary era, large ice sheets and glaciers have continually advanced and retreated. It is perfectly normal for glaciers to grow and melt during the ice ages. We see the same thing happening today. It is also perfectly normal for large chunks of ice to break off and fall into the sea. A large rift has recently appeared on the Larsen C ice shelf in Antarctica. This process is called calving and occurs when the ice becomes eroded by the sea. This entirely natural phenomenon has been seized upon as evidence of a global warming, but in fact calving has occurred throughout the ice ages.

In 2017 a large chunk of the Antarctic ice shelf calved and floated away. This did not cause a rise in the sea level however, since the ice is already floating on the water. According to NASA Goddard Institute for Space Studies the Larsen ice calving event contributed to a mere 0.1 mm rise in sea level! When you float ice cubes in a drink and the ice melts, your drink does not overflow!

Since the Antarctic ice shelves are floating in water they will not raise sea levels if they melt. The calving process is a normal sign of a healthy ice sheet that periodically sheds its load in order to maintain homeostasis. This is because frequent heavy snow falls will cause the ice sheet to become heavy and unbalanced and therefore a calving has to take place.

Although we are in the midst of the Quaternary Ice Age, we are fortunate to be living in one of the interglacials of this formidable ice age. The interglacials of the Quaternary ice age generally have a length of between 10-12,000 years. Ice ages have shaped human civilisation. The gift of fertile soil to farmers is the legacy of the last ice age.

The breadbaskets in America, and the nutrient rich soil of Great Britain, were enhanced when the ice retreated. Human civilisation has been flourishing for nearly twelve thousand years in a balmy interglacial. What does this mean exactly? It means that we could be at the very end of our warm weather interglacial. We are now hurtling towards a major glaciation event and a recurrence of the great ice age!

The present geologic eon is the Phanerozoic Eon, which spans the past 545 million years. This Eon encompasses twelve periods. These lengthy periods spanning millions of years each are known as; the Cambrian, the Ordovician, the Silurian, the Devonian, the Carboniferous, the Permian, the Triassic, Jurassic, the Cretaceous, Paleogne (emcompassing the Paleocene, the Eocene and the Oligocene) the Neogene, and finally the Quarternary Period.

Readers may be familiar with the Triassic, Jurassic period as being the time of the dinosaurs. The fertile Carboniferous period had warm tropical swamps and huge trees that created many of the carbon coal deposits buried deep underground that are mined today. During our Phanerozoic Eon, planet Earth has endured several frozen climate cycles, some lasting 50 to 90 million years. The present day Quarternary ice age period is the shortest geologic period in the Phanerozoic Eon.

The Quartenary divides into the frozen Pleistocene ice age, and the present day Holocene interglacial which has seen mankind flourish with its favourable weather conditions. This benevolent interlude of interglacial warmth may be about to change as the Holocene is due to end with the dawn of a new ice age. Why did the climate become so much colder during the last 2.5 million years?

One of the reasons for this colder climate is that warm ocean currents became blocked when South America moved towards joined North America. The Panamanian Isthmus then became a land bridge. This bridge then disrupted and blocked the warm Pacific Ocean waters, thereby preventing them from mixing with the much colder Atlantic Ocean waters. This Panamanian Isthmus then gave rise to the thermohaline circulation that is part of the Gulf Stream. Because of this geological formation of the land bridge, ice was able to accumulate at the Arctic and has remained there ever since. This example shows how a tectonic structural change in land mass can affect the climate (Oceanus 2017).

The Antarctic had already been ice capped for millions of years after Gondwanaland broke up. Gondwana was originally part of the supercontinent Pangaea. The land mass started to separate during the Jurassic period. When this super continent divided the Antarctic region was left isolated from warmer ocean currents and became frozen. The Arctic North Pole then also froze up after the Panamanian land bridge was formed.

The opening and closing of seaways has a profound effect on the heat that is transported by ocean waters. A slight tilt of the Earth on its axis 2.5 million years ago might have resulted in less sunlight in the Northern Hemisphere. So now both Poles had permanent ice caps that have lasted to the present day. Despite fears that they will melt, this is highly unlikely due to the aforementioned factors.

The ice ages that form part of the present ongoing ice age, the Quarternary, usually last for around 90-100, 000 years. This figure is based on the research of husband and wife team John and Katherine Imbrie who have written many books on the history of Earth`s great ice ages. At the grand finale of an ice age there is a rapid transition to an interglacial. Indeed prehistoric man living in the freezing ice age conditions could have literally woken up one day to find that the ice was melted.

Within a decade or so our surprised ancestors could have unexpectedly found themselves living in a balmy paradise! So it seems that a transition from an ice age to an interglacial can happen very rapidly. Therefore one could surmise that the converse could occur, and that an ice age may arrive just as rapidly within a few decades. Indeed the geologic temperature graphs show a rapid descent into an ice age has taken place previously.

The Earth has always existed in one of three conditions. These three conditions are a greenhouse, an icehouse and a snowball Earth. The last two million years have been mostly very cold on our planet with ice caps covering much of Europe, Asia, North America and two thirds of the British Isles. The "Icehouses" are about 5°C colder than the four warmer epochs known as "Greenhouses". Millions of years ago dinosaurs basked in tropical sunshine in England.

Alligators roamed as far north as present day Alaska! There was no ice present at either of the Poles or indeed anywhere on the Earth. The Earth was then in a *real green house* state. A proper greenhouse in scientific terms is when there is not a trace of snow or ice on planet Earth. Our planet has since passed into an *ice house* state. The present ice house state is the reason that we presently have plentiful snow and ice at both North and South Poles.
We are already living in an ice house condition and soon it will get much colder.

Earth has also been in a snowball condition where the entire globe was covered in snow and ice. "Snowball" Earth condition is thankfully very rare. The last snowball Earth event is known as the Cryogenian. The snowball began 850 million years ago and globally temperatures were minus 12 degrees below freezing. Land froze over and seas turned to slushy ice. Snowball Earth is very rare thankfully, as were it to occur today, life on Earth would all but be wiped out. It is an amazing concept that the Equator was once covered in ice!

A 2010 study by Harvard scientists shows that the Earth did indeed freeze over like a snowball (MacDonald 2010). Imagine a scenario where the tropical equatorial zones are frozen and the oceans are nearly frozen. The evidence from geologic sediments does indeed point to the Equator once having Arctic conditions. Resilient life forms persisted in the cold oceans. The snowball finally melted 636 million years ago. When the snowball finally thawed life exploded across the planet in an evolutionary burst.

There are many theories as to how this snowball event happened. One of these theories points out that the sun was much weaker at that time. The sun has been shown to regularly fluctuate in output and this variability and its effect on climate will be discussed in this book. The sun may also have influenced the melting of the snowball as the Earth`s orbital tilt changed so allowing more warmth from the sun`s rays to penetrate the ice. Underwater volcanic eruptions may have helped trigger the great melting of the giant snowball.

At the present time we are actually living in an icehouse condition. This fact may seem at odds with all of the media attention that is focussing on a runaway global warming greenhouse scenario for our precious planet. Currently global temperatures are just 5 degrees Celsius above the previous global temperatures during the last ice age. Therefore we are not so far off an ice age already.

During the past two million year ice age freeze of the Northern Hemisphere, the climate has been glacial for 90 per cent of the time. This ice age freeze has been punctuated by short warm periods totalling a mere ten per cent of the overall time period. Therefore ninety per cent of the last two million years on planet Earth has been extremely cold and very thick ice sheets have amassed. The remaining ten per cent of the warmer climatic periods is therefore atypical of the climatic norm for our planet.

We are presently living in one of these atypical warmer periods. Worryingly this warm interglacial that we enjoy living in is predicted to end soon if the past data is correct. Those who hope that carbon dioxide levels have risen sufficiently to stave off another ice age may be proved wrong. Northern Hemisphere winters have recently accumulated similar snow cover to that of the previous ice ages.

This extensive snow cover is clearly shown from photographs taken from the air during recent snowy winters. The warning signs are all pointing to the onset of the expected ice age and we will be powerless to prevent it! The Northern Hemisphere currently experiences heavy snow fall as we are still in the present "icehouse". The reasons for an ice house condition have been speculated to be due to Earth`s passage through a dense matter spiral arm of our Milky Way or it may be a result of structural tectonic break ups on our planet.

Either way the present ice house conditions that we are living in could be about to get a whole lot worse! When our ice age finally arrives with a vengeance we could see a decimation of the population from hypothermia and starvation. The reasons why Earth has been so very cold for the last two million years are not fully known, though there is a theory of that tectonic upheaval and the Panamanian Land Bridge may have triggered the cold Earth condition.

A supercontinent known as Pangea dominated Earth for 160 million years. Pangea later broke apart into the continents of Africa and America. This separation is conjectured to have somehow caused a shift in the climate. There is scientific evidence that proves that climate was drastically altered as Pangea broke up into large chunks of land masses. When the continents separated a change resulted in the flow of the oceanic currents and winds.

The scientific theory is known as Continental Drift and was created by Alfred Wegener to explain how the continents shifted Earth's surface. This drifting of continents affected aspects such as climate and rock formations. The outcome of Pangea separating was that the world became cooler as the oceans currents became displaced and ice caps were formed. The Himalayas rose up later in another tectonic upheaval. This mountain formation made the planet cooler by changing wind patterns, according to the Centre for Ice and Climate at the Neils Bohr institute.

Our ever changing planet Earth has experienced five major ice ages that have been named and documented. We are presently in a fifth ice age known as the Quaternary ice age. Prior to the present Quarternary Glaciation there have been four other long lasting ice ages. Before our present ice age there was the Karoo ice age that took place during the Carboniferous and Permian eras 360 to 260 million years ago.

The Karoo ice age was preceded by the Andean –Saharan ice age during the Ordovician and Silurian epochs. This ice age took place 450 to 420 million years ago. The Karoo ice age was preceded by the Cryogenian ice age 850 to 635 million years ago. As its name suggest the Cryogenian ice age was very severely cold. It is theorised that a complete snowball Earth glaciation occurred during the Cryogenian (Kopp et al 2005). It is hypothesised that the snowball finally melted due to a spate of underwater vulcanism that would have warmed the seas. The melting of the giant snowball may have then triggered the so called great oxygenation event that caused an explosion of life on our lifeless frozen planet.

The first recorded ice age is known as the Huronian Ice Age. This took place from 2.4 to 2.1 billion years ago. This ice age led to mass extinction of the primitive bacterial cellular life forms that were emerging. It is thought that the sun shone dimly at that time, making Earth icy cold. Interestingly Kopp`s team discovered that the Earth`s atmosphere became colder when carbon dioxide and water emerged during this time.

This is an unusual finding since both water vapour and carbon dioxide are widely considered to be green house gases! This made the atmosphere thinner and it lost heat. This ice age was triggered by the great oxygenation event that resulted in the formation of carbon dioxide and water which replaced the methane atmosphere. This cooling event is an interesting fact since it is held that carbon dioxide is an important warming greenhouse gas.

Ice core researchers at the Neils Bohr Institute in Denmark discovered that the Earth has been getting gradually cooler over the past 40 to 65 million years. This long term ice core data does help imbue a sense of perspective regarding long term climate change. We may perceive our climate to be warming in the present day but the fact remains that Earth has been slowly cooling down over the last 40 to 65 million years!

Danish ice core researchers think that this gradual cooling was the result of the continents drifting and altering the temperature somehow by amplifying Earth`s orbital radiation influx. These orbital influences are sometimes covered by the term Milankovitch cycles that are named after the Serbian mathematician Milutin Milankovitch.

The Neils Bohr institute also discovered that during the past 40 million years levels of carbon dioxide have also been falling. Levels of carbon dioxide were once as high as 2000 parts per million (ppm). In contrast today`s levels of carbon dioxide may be rising to a level of 400ppm. Many concerns have been voiced by this high reading that is blamed on our industrial activities. One can observe that in the long term paleoclimatic history of our planet there have been times when carbon dioxide levels were five times greater than they are in the twenty first century.

Today there is a lot of anxiety about the fact that atmospheric levels of carbon dioxide have been climbing steadily since our industrial revolution in Great Britain. The carbon dioxide levels seem to be slowly creeping up and have now reached the dizzying heights of 400 parts per million. This relatively modest rise is causing panic among some environmental activists and there have been many protests about climate change. Therefore it is important to take a long term perspective of the Earth and to appreciate that this level of 400 parts per million is well within the range of what is normal for our planet.

The long term trend is for a cooling Earth. Our planet has steadily cooled over the last 40 million years. This is an important statistic to remember. 40 million years is a long time for a downward trend of global temperature. If the climate became briefly warmer during recent decades this temperature blip would appear as insignificant noise on a long term temperature graph of our planet`s paleoclimate.

The graph below spans millions of years. There is a linear downward cooling trajectory. We are at the end of the axis in the Pleistocene epoch at the coolest part in Earth`s history.

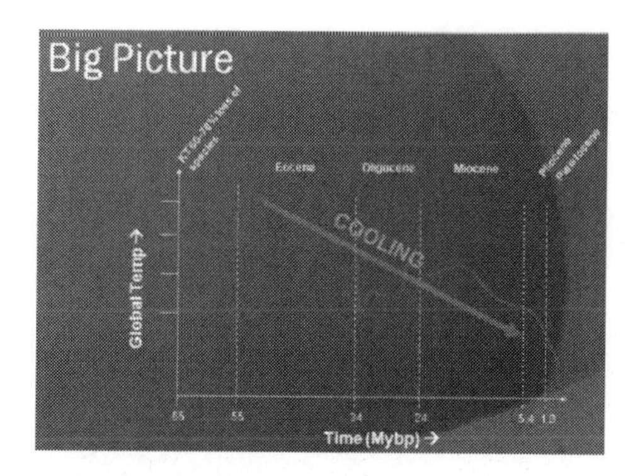

Credit Dandebat/DK

The Pleistocene/ Quarternary epoch started 2.6 million years ago and has consisted of several ice ages. Thus we have an inexorably cooling Earth, though we are presently enjoying an interlude of milder warmer conditions in the Northern Hemisphere. It is important to get the big picture about climate on Earth. The long term trend shows a cooling which may become dangerous to humanity if it continues.

The pattern of ice ages that we presently experience commenced 40 million years ago and seems remarkably durable. It may seem baffling to talk of the prevailing pattern of ice ages when we are enjoying the relatively mild weather of our era. This is because we are still enjoying our interglacial epoch known as the Holocene. This book will explain why our interglacial may be about to end, and thereby plunging us into an instant ice age of unimaginable severity.

This well known graph of the Pleistocene ice age shows the small areas above the base line which depict the warm interglacials. We presently live in the last circled area above the base line. For the last three million years the majority of the time our planet was in the cold ice ages depicted as shaded areas below the base line. This graph clearly shows that we are presently in an anomalous period of warmth which may soon be about to end. The precipitous fall into ice age conditions from the peak of an interglacial should give us cause for concern.

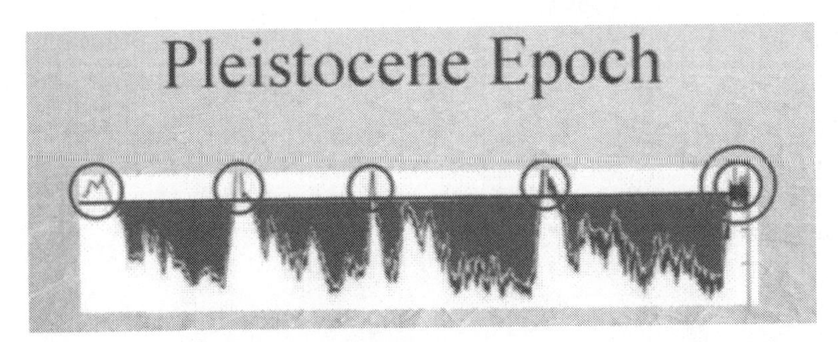

Credit: Seawapa org.

The recent long term paleoclimatic trend is for a warm interglacial to last for approximately 11 to 12 thousand years. Mankind has now been enjoying life in the balmy conditions of this interglacial for 11, 726 years to give a more precise measurement. Therefore we could be about to exit our warm interglacial known as the Holocene. The idea that our greenhouse emissions may stave off this ice age is wishful thinking unfortunately. In past geologic epochs such as the Ordovician, there have been levels of carbon dioxide as high as 3000 ppm and yet the Earth still endured terrible ice ages.

Geological evidence of the last ice age comes from the large displaced rock boulders called erratics. These gigantic boulders have been found all over America. There is even a large erratic boulder in Central Park, New York. The large boulders were pushed south by the ice sheets. Geologists studying the ice ages have finally explained the mystery of these large boulders called erratics which are found in unusual locations all over America.

The Earth has many strange quirks regarding climate and so we cannot say this departure from the Holocene will definitely occur any day now with total confidence. It is possible that our interglacial might last a lot longer than is the norm. A longer interglacial has happened in the very distant past. However since the million years long trend shows a cooling of the Earth, this might be overly optimistic. The probabilities are that this ice age coming soon will be the worst ice age on record and may even wipe us out as Earth is inexorably cooling down.

The ice ages used to thaw out every 41 million years. Then this pattern changed to a thawing out once every 100, 000 years. This change to a longer ice age implies an insidious cooling. The Earth is now on an inexorable cooling trend in the paleoclimatic context. Therefore this interglacial may not be late in ending as is optimistically hoped by climate scientists. An ice age could commence during this century or even in a few years time!

There once existed an interglacial that lasted 28 thousand years according to data from the European Project for Ice Coring in Antarctica known by the acronym EPICA. This might give rise to hope that we may now be an interglacial lasting longer than the general duration of around 11, 500years. This is a complex subject and some ice core specialists think that our present glacial may be longer than 11, 700 years. However it would be sensible to prepare ourselves in case the warm Holocene comes to an abrupt end.

When discussing the ice ages it is important to note that the Antarctic and the Arctic behave differently. Whereas the Antarctic has become locked into a more or less fixed cycle of continuous ice, the Northern Hemisphere behaves differently. In the Northern Hemisphere over millions of years, a pattern has developed that differs from that of the South Pole.

The Northern ice sheets and glaciers have developed a pattern of continual advance and retreat which differs from the relatively static pattern of the Southern ice sheets. So the next time you see a time lapse photo of Arctic ice retreating, rest assured that this is how the Arctic ice sheet behaves. The main ice advance cycles have occurred at 21, 41 and 100 ky cycles.

There is often speculation as to how the massive ice sheets in North America could have melted. The thawing of the American Laurentide ice sheet might be explained by the sinking of rock under the colossal weight of the North American ice sheet. After several ice accumulation cycles the Earth's crust sagged so far that the ice's surface was at a low enough altitude to melt in summer.

This is known as isostatic rebound. Geologist Peter Clark of Oregon State University speculates that the melting of ice could have been triggered because the ice sheets had reached such a size that they had become unstable and were ready to go. However isostatic rebound does not explain the exact time scales of the ice ages.

The innovative American geologist, Randall Carlson, thinks that the energy needed to melt the ancient Laurentide ice sheets would be in the region of that produced by thousands of megaton nuclear blasts! So clearly he is not a big fan of the isostatic rebound theory. Carlson thinks that a super massive event such as a massive solar flare storm or an asteroid impact, maybe even a huge volcanic eruption would be needed to melt this huge ice sheet.

Carlson has a lot of original ideas and he speculates that 11, 600 years ago there was a great flood that destroyed the mythical civilisation of Atlantis that is now below the sea. Atlantis was a Greek island that Plato wrote of 11, 600 years ago. It is also the time of the ice thaw and melt flood that ended the ice age. Could the ancient myth of Atlantis possibly be true? Carlson does not subscribe to the popular idea that rising carbon dioxide levels would have been able to melt the very thick frozen ice sheets.

In truth no one really understands how these massive sheets could have melted so suddenly and this is very worrying as there may be cataclysmic events that we are not prepared for. Perhaps the Sun woke up? The popular theory of the Sun as a variable output star implies that the Sun could go to sleep during an ice age, in a metaphorical sense. The last ice age may have ended with a frightening bang but we do not know yet how the next ice age will begin. It could begin with a gradual accumulation of snow or with a snowblitz of unprecedented intensity.

The Northern Hemisphere is densely populated as the greater land masses are here. A pattern of heavy snowfall and advancing ice sheets therefore makes the Northern Hemisphere more prone to an ice age. The ice core data implies that the Northern Hemisphere has more to fear from a sudden increase in ice cover as this where the greatest land mass and population lies.

Ice core studies show that prior to the last 40 million years the warm interglacials lasted for much longer time periods. Then something happened to switch this pattern and the warm interglacials began to become shorter. We are presently living in one of these warmer interglacial times of this very severe ice age.

Some scientists think that the Northern Hemisphere may be more vulnerable to small changes in insolation or sunlight reaching the Earth`s surface (Neils Bohr Ice and Climate Unit). This is corroborated by ice core records and the present day phenomenon of Arctic amplification. Arctic amplification is a process whereby small differences in insolation lead to a disproportionately bigger effect in the poles. The idea that a small change can lead to a larger non linear effect was demonstrated by chaos theory. This concept sometimes known as the butterfly effect will be discussed later in relation to the Earth`s climate.

Returning to the present Quarternary ice age, it seems that our present interglacial may soon be ending. The incontrovertible geologic record proves that most interglacials last for around 11, 500 years. This deduction is based on past geologic records for the Quarternary which is also known as the Pleistocene era. For the past two and a half million years the Quarternary ice age has moved in and out of glacial conditions.

Evidence for past ice ages has come from geological phenomena such as glacial sedimentary till and displaced boulders known as erratics. Geologists found huge boulders which they named erratic.

These gigantic rocks had been pushed for miles by moving ice sheets. Ice age landslide debris also scratched the bedrock.

The end of the last ice age may have coincided with a major catastrophe. Earth has suffered at least five major catastrophic extinction events during its turbulent history including the Cretaceous-Tertiary extinction event 65 million years ago that wiped out the dinosaurs. This was preceded by the Permian mass extinction 248 million years ago that wiped out 96% of all species on Earth.

Our catastrophe free interglacial, the Holocene, is unusually calm. While it has been beneficial for humans there is evidence that many species are becoming extinct as a result of unfettered human activity. This had led scientists to propose that our Holocene be considered as the sixth major extinction event in Earth`s history. The present rate of extinction is around 140, 000 species per year. The Holocene extinction started at the end of the Pleistocene.

At the finale of the last Pleistocene ice age the megafauna or giant animals died out. Gigantic sloths the size of a small truck suddenly died out. Most of the giant sloths, bears, sabre tooth tigers, mammoths and hippos suddenly became extinct 12000 years ago. Perhaps these megafauna were hunted by early humans or perhaps an impact wiped them out. We do not know what caused this Quarternary extinction event that occurred at the end of the last ice age.

It has been speculated that farting mega -herbivores would have contributed to the atmosphere`s greenhouse gas methane! Therefore their extinction and consequent lack of farting warming emissions may have contributed to the subsequent Younger Dryas cooling event! However other studies show a fall rather than a rise in greenhouse gases during the Pleistocene ice age and so this academic area is not settled.

To avoid confusion it should be noted that the terms Pleistocene and Quarternary ice age are both used interchangeably to describe the same ice age. During the Pleistocene-Quarternary ice age epoch the melting events used to take place every 41 thousand years. Then the pattern changed to a melting event every 100, 000 years. So the ice ages now lasted around 90-100, 000 years followed by a melt into an interglacial. This longer 100 000 year thaw cycle is the one we presently live in.

The fact that the pattern shifted to a longer time period has puzzled many glaciologists. It may even indicate that the Earth is gradually getting cooler since the ice ages are now lasting for a much longer duration. What caused the melting events to start taking place at further apart time intervals? This conundrum has been called the 100,000 year problem. It is also important to make a mental note of this 100, 000 year dilemma because it calls into question the theory of the Milankovitch cycles and orbital timescales which will be discussed later.

The transition from the 41 thousand year intervals to the longer 100, 000 year time interval is known as the Mid Pleistocene Transition or MPT. This transition took place around 800, 000 years ago during the icy Pleistocene epoch. Why is any of this important I sense my readers are asking? This transition from 41 to 100 thousand years matters because it implies that in the long geologic time frame our planet is getting gradually cooler. Some geophysicists think that the Earth`s core is also getting cooler.

Might there even be a link with a cooling inner core? The Earth`s molten core is 4.6 billion years old and is no longer a teenager. The swirling outer core generates a magnetic field via the rotational coriolis force. However as will be discussed in the chapter on the magnetic field, there are ominous signs of a magnetic field collapse. A slow cooling trend might explain the Mid Pleistocene Transition. The Earth is no longer thawing out every 41,000 years because it is slowly getting colder.

When the Earth first formed there was plenty of heat in its core and this in turn gave rise to our life giving magnetic field. If the Earth is cooling down this could be a worrying portent. There could even be a cosmic connection causing the insidious cooling such as a dense dusty spiral arm of our Milky Way that is serving to block some of the solar input.

The Earth periodically dips in and out of ice ages and scientists have looked at orbital planetary mechanical factors to explain this. These orbital and axial tilt influences were first discovered by James Croll who was a Scottish scientist born in 1821.

Such mechanical influences affect the levels of incoming solar radiation as sometimes the Earth`s surface is further away from the warming sun. Croll developed a theory of climate changes on Earth that were caused by changes in the Earth`s orbit. Sometimes the Earth follows a circular orbit and sometimes the orbit becomes stretched into an ellipse. Croll believed that when Earth`s orbital eccentricity was high the decrease in sunlight would favour an ice age.

These ideas were further refined by a Serbian scientist called Milutin Milankovitch. Nowadays not everyone is familiar with the celestial Milankovitch cycles that are deemed to affect our climate. The popular press would have us believe that it is only carbon dioxide that affects our climate. The three Milankovitch cycles comprise of axial tilt, precession and orbital eccentricity.

Axial tilt is also known as obliquity. The tilt of Earth's axis fluctuates by around two degrees in a 41,000-year cycle. Moreover, Earth's axis gyrates in a cycle of 26,000 years, much like a spinning top. Finally Earth's elliptical orbit around the Sun changes in a cycle of around 100,000 years. The axial tilt of our planet is now at 23.5 degrees and this tilt gives rise to our four seasons.

Precession is the term used to denote the direction of the Earth`s axis. Today we face northwards facing our bright shining North Star also known as Polaris. However around 13, 000 years ago the North star would have been Vega as the Earth`s axis was tilted in a different direction. These concept might seem difficult to understand so one suggestion is to imagine a toy spinning top. It wobbles on its axis and the tilt changes as it slows. This is how the Earth spins and wobbles on its axis.

Seth Chandler was an American astronomer who discovered that Earth is wobbling like a spinning top. Chandler said that the Earth "wobbles like a top" whenever our planet slows down in its rotation as it has done in recent years. According to NASA, "the track of this spin began to slow down very slightly about January 18, 2006. Since then, we've had a series of extremely harsh winter seasons in both hemispheres .If Chandlers` wobble of the planet continues, it is possible that we will eventually see at least a new Little Ice Age, maybe even a great ice age like the one that ended approximately 11,700 years ago.

Precession cycles might affect our seasons here on Earth. The precession cycle may have been responsible for the climatic optimum of our Holocene interglacial. This climatic optimum occurred 7000 years ago and resulted in warm Northern Hemisphere summers. The summers were much warmer than our present day summers during the mid- point of our interglacial. A precession cycle lasts for 23,000 years. An interglacial lasts approximately half of a precession cycle which is 11, 500 years.

The Northern Hemisphere is now tilting away from the sun and so our summers are cooler than they could be. Indeed temperatures in our interglacial have been gradually decreasing for the last 6000 years after the Holocene Climatic Optimum. This fact is very important in our discussion of an ice age event in Britain as it indicates that we may be slowly gearing up for our overdue ice age.

It is thought that a reduced summer insolation is one of the key drivers for a Northern Hemisphere ice age. Therefore at this present time we are in fact in the correct position for an ice age to occur in the North. It is theorised that cooler summers will result in less of the snow melting on higher ground and therefore eventually a snowy albedo will lead to further cooling. Land masses in northern latitudes primarily dictate the onset of ice ages, and so this is where our attention should be directed.

It is theorised that Ice advance cycles in the Northern Hemisphere are affected more by the orbital and axial tilt cycles than the Southern Hemisphere due to a larger land mass. When the large land masses become snow covered there is a feedback loop whereby the white snow reflects sunlight back into space. This is known as the albedo effect and it serves to further cool the planet until one day the snow and ice no longer melt. Further snowfalls add to the existing snow and the layers pile up until there is a mile thick ice sheet.

We are presently in this mode of cooler summers and warmer winters phase of the Milankovitch cycle that is thought to be a prerequisite for an ice age in the Northern Hemisphere. The theory is that snow accumulations will start to settle as the summer is not warm enough to melt it. In other word it is the small redistribution of the heat over the seasons with colder summers that tip Earth into an ice age.

How could such a small seasonal difference in incoming warmth make such a difference to our global climate? Milankovitch felt that it was the variable of insolation striking at the latitude of 65 degrees north that affected the ice ages. Cooler summers result in less ice melting and gradually the albedo accrues. The ice core researchers, Hays, Imbrie and Shackleton, concur that the 100,000 year ice age cycle is in phase with orbital eccentricity. This assumption only works on the basis of non linearity however (Imbrie 1992).

Eccentricity is a measure of how circular a curve is. Therefore the parameter of eccentricity describes the Earth`s circular to slightly elliptical orbit around the sun. This slight non circular orbit means that solar insolation levels vary slightly. Earth`s orbital eccentricity is very small however at 0.02 deviation and the orbit is nearly circular. This is important for understanding ice ages as levels of solar insolation need to decrease to initiate an ice age. It is amazing how such a small deviation in orbit could make such a vast difference to our climate.

This 100, 000 year eccentricity cycle seems to have a strong effect on ice ages. Eccentricity is the only Milankovitch cycle that affects the total energy received from the sun. However since the orbit is only ever slightly eccentric the radiative transfer is of the order of -+ 0.2 per cent change in solar insolation. At the present time Earth is at the minimum phase of the 100,000 year eccentricity cycle. In other words we are in a circular orbit rather than an elliptical orbit. Therefore this parameter is not good enough evidence for our overdue ice age.

Many theorists are not at all happy with the Milankovitch theory and find discrepancies between ice ages and the orbital factors. For example there are big discrepancies in the theory and the tropical climate of the dinosaur era in the Jurassic. This theory also does not explain why the climate on Earth keeps getting cooler in the long term trajectory. Earth has been cooling down over the last 40 to 65 million years. Therefore it is at best a theory and there must be other major parameters involved in ice ages.

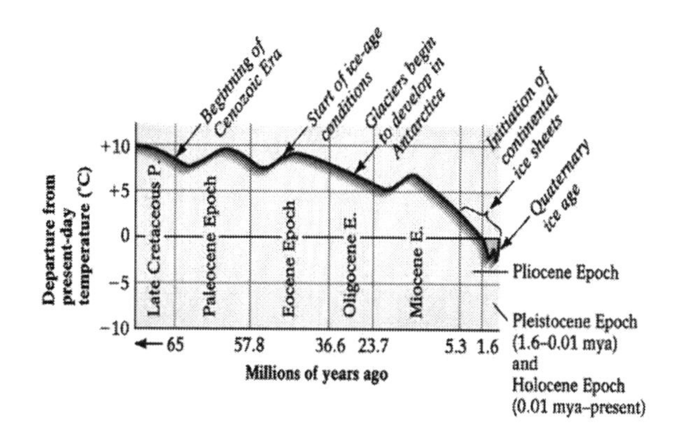

The graph shows a decline in temperature over the last 65 million years, Credit: Dandebat/dk

65 million years ago a massive asteroid is thought to have impacted on Earth according to a discovery made by Walter and Luis Alvarez in 1980. Is it possible that the KT asteroid impact during the dinosaur era in has knocked the Earth into a cooler axial tilt? This is my own idea and since I am not an astrophysicist it could be erroneous! The Mexican Yucatan Peninsula Chicxulub crater say geologists, provides firm evidence of a huge impact. Geologists have found a defined layer of extra terrestrial iridium at the time of the impact that reaches around the entire globe.

Greenland has been the favoured location for many ice core climate studies. The temperature of the Earth is past eras has been gauged by measured fluctuations in oxygen isotope rations. Oxygen comes in three isotopes with differing numbers of neutrons. Oxygen-16 and oxygen-18 can tell us a lot about the climate in centuries past. Oxygen-16 is correlated with colder icier climatic conditions. Glaciologists have discovered violent isotopic changes that indicate dramatic climate fluctuations must have taken place. The climate must have become colder by 20 degrees Celsius within a very short space of time amounting to mere decades. An isotopic study by glaciologist Professor Dowdeswell in 1995 indicates that the Earth could have plunged almost instantly into ice age conditions.

Returning to Milankovitch theories many scientists find it hard to believe that a mere redistribution of the exact same amount of heat could somehow lead to an ice age. Lest we take this theory as gospel there are many scientists who think the theory is over rated and may even be incorrect. For instance a shift has occurred in the ice age cycle. The glacial cycles became longer lasting 100, 000 years instead of 41, 000 years. The recent 100, 000 year long ice age cycle does *not* correlate well with the Milankovitch predictions of insolation levels either. Therefore we need to look further afield for ice age explanations.

One thing to consider is that our solar system is not stationary but is travelling at 515, 000 miles per hour through the Milky Way. The sun takes 225 million years to move around the entire galaxy and this is called a galactic year. The Milky Way is also moving incredibly fast through space. Indeed it is on a collision course with our nearest neighbour the Andromeda Galaxy according to the Hubble telescope. One day the two galaxies will collide and merge tossing planets into space in a gigantic upheaval. The new galaxy will be elliptical in shape and contain many black holes and bright stars.

The telescopes show the Orion Nebula which is 100 light years away from us. This region is the birthplace of stars. In the past planet Earth could have drifted in the midst of such stellar nurseries leading to cataclysmic impacts. A super nova explosion is caused by the death of a star. A supernova only 30 light years from Earth would bombard us with x-rays, destroy the ozone shield and lead to a mass extinction on Earth. We tend to be cuckooned in a cosy bubble believing that Earth is safe in its trajectory through turbulent outer space but we are wrong. Space is full of hidden dangers that may cause mass extinctions or even trigger ice ages.

Richard Muller is a geoscientist at Berkeley, California who favours a space dust theory for ice ages. Muller is famous for his Nemesis Death Star theory which hypothesises that an invisible red dwarf orbiting our sun may periodically dislodge comets towards Earth. Even if there is not a Nemesis there may be other forces that can dislodge comets. For instance, an angular effect of the galactic gravity plane could cause orbital perturbations of outer Solar System objects (Kenyon et al 2004).

Muller predicted that if anybody could ever measure the variations in dust, they would see a 100,000-year cycle that would match the cycle of the ice ages. The famous astronomer Fred Hoyle who coined the term "Big Bang" also believed that cosmic space dust or diamond dust in the upper atmosphere could reflect sunlight and lead to an ice age. A geochemist at Caltech looking at sea-floor sediment cores found increases in dust at intervals that matched those of glacials.

The space diamond dust reflecting dust theory fits the climate history data very well, and doesn't have the causality problem and need for feedback variables of the Croll/Milankovitch theory. Muller speculates that the incoming space dust affects the electrical currents that help the seeding of clouds. This space dust may be a result of passing through a spiral arm of the Milky Way every 100, 000 years. Certainly this theory overcomes the many discrepancies found in the Milankovitch model for ice ages.

Evidence to support the cosmic space dust ice age theory is found in ocean sediments. A comet may have broken up 3 million years ago leading to this dust residue and initiating the Quarternary ice age.. It is well established that ice ages were very dusty and dust deposition increased two and half times (Higgins et al 2002). Regardless of which theory is correct we are overdue an expected ice age.

Perhaps the drier colder climate and the strong cold winds may have played a part in whipping up dust storms. Scientists want to find out exactly where this dust came from. It is possible that it is extra terrestrial and might be caused by small meteorites breaking up. The dust contains iron which is beneficial for ocean life. Marine ecology does very well in ice ages. If an ice arrives it may give the oceans time to heal. It is possible that Earth was bombarded with tiny meteors or minute diamonds giving rise to the extra dust.

Maybe large scale volcanic eruptions occurred during the ice age adding to the extra dust. It is important that researchers find the origins of this dust so we know what to expect during our next ice age. There are substantial measured increases of interplanetary dust arriving during the 100, 000 year long ice ages.

This finding of cosmic dust on the sea bed implies that Earth might pass through an interplanetary dust cloud during an ice age. Professor Sean Higgins and his team took soil samples from the Ontong Java plateau. The team think that extra dust might be the driver for the ice ages. The team found no evidence to support the theory of Milankovitch cycles. It seems that many holes can be found in the orbital, axial tilt parameters and therefore one should not take this theory as gospel. Some scientists are now proclaiming that an ice age cannot possibly arrive as the Milankovitch cycle is not receiving sufficiently low levels of insolation.

A discussion of innovative theories would not be complete without mentioning the mysterious and legendary Long Count calendar predictions of the ancient Mayan astronomers. The Mayans were an ancient Mexican civilisation who made astonishingly accurate computations of astronomical cycles. The Mayans understood the cosmic principles of alignment and precession.

The Mayans were advanced mathematicians and had precise astronomical calendars. The Mayan calendar calculated that a new cycle would commence on Earth starting from the winter solstice in 2012. It is just possible that this new age predicted by the Mayan astronomers might refer to the onset of a major ice age. This is a feasible hypothesis since ice ages are influenced by astronomical variables.

There has been some speculation that the black hole at the centre of our galaxy may exert an unknown force that affects our planet. Astronomers now believe that super massive black holes lie at the centre of all galaxies including our own Milky Way. Our black hole is situated in a region known as Sagittarius A. In 2015 NASA observed a strong x-ray signal from Sagittarius A that was 400 times more powerful than usual signals. They theorised that it was caused by an asteroid being sucked into the black hole. This event proves that our supermassive black hole is one of the highly active black holes in our universe as some are more quiescent.

The Mayans thought that an alignment with the centre of the galactic plane was of great significance. The galactic plane is where the black hole is sited. It is speculated that the galactic plane is therefore a region of greater electromagnetic and gravitational disturbance. However some astronomers say that Earth has only passed through the galactic equator and that this is not the same thing as the galactic plane, so there is really nothing to worry about.

It is just possible that the Mayans knew from their advanced astronomical mathematical calculations that a new cosmological era would arrive on Earth around the year 2012. The new cycle would commence on the December solstice when the sun would be in perfect alignment with the centre of the Milky Way. Interestingly a Mayan prediction for this date states that "a rebirth will occur for the entire world when there is a future union with the December 21 winter solstice solar lord Great Mother`s galactic heart".

The heart seemingly refers to our hypothesised black hole at the centre of our galaxy. Did the ancient Mayans even know of a black hole? By tracking precession the Mayans were able to compute that this solstice date marked the coming of a rare alignment. A rebirth of Mother Earth could certainly take place as the Mayans predicted. The ice age would give our planet a chance to recover from the terrible damage that is being inflicted by us. An ice age may be bad news for humans but it is certainly good news for our suffering planet.

Intriguingly after the year 2012 the Mayans gave up with their calendars. Perhaps they thought that civilisation on Earth would soon be finished thereafter and there would be no more need for a new calendar? Ominously the Mayans foretold that large numbers of “souls will be harvested” which implies that a mass extinction might occur soon. Thankfully we are all still here safe and sound after the winter solstice of 2012. The Mayans may have been making a symbolic prediction using terms they were familiar with, that do not make complete astronomical sense today. NASA sought to reassure us that the solstice would not mark the end of the world and that nothing untoward would occur!

However we cannot afford to be complacent as there may be a cosmological event that is starting to unfold such as a comet or meteor being dislodged from the Asteroid Belt, the Kuiper Belt or the Oort cloud. The Kuiper belt is a large ring of observable asteroids near Neptune and includes the planet Pluto. The Oort cloud is much farther away and composed of icy planetesimals and comets. This icy cloud is a hypothesis used to explain the sudden arrival of long period comets.

The main asteroid belt lies between Mars and Jupiter. Its largest asteroids are Ceres, Vesta, Pallas and Hygiea. Indeed it seems highly likely that an asteroid might be dislodged at some point since there are so many circling us. The Dark Rift of the galaxy is thought to be dark due to a high mass of cosmic material such as meteors. This area extends from the outer solar system to the Sagittarius arm of the Milky Way. It seems that somewhere out in space is an asteroid, comet or meteor with our name on it!

In the year 2000 there was a sensational headline In the Alaskan Weekly World News that a geophysicist had discovered that the Earth`s rotation is slowing down. Dr. Kopeski caused minor panic when he announced that if the Earth`s rotation continued to slow there could be one half of the globe in total darkness with the other half in perpetual daylight! In days gone by, a day on Earth once last only 13 hours. Over time the days have become longer as the rotation slowed down.

In December 1998 something strange was observed by astrophysicists as the Earth`s smooth axial wobble on its axis was deflected. It was widely reported that the so called Chandler`s wobble had stopped though it later recovered. This could be a sign that chaos or non linearity is starting to develop in our axis. The MIT Technology review reported in 2009 that the wobble in Earth`s axis had changed phase by 180 degrees in 2005.Perhaps the wobble is starting to become chaotic?

Chaos theory demonstrates how a small initial change in a parameter can later magnify and turn into a huge change that is out of all proportion to the initial parameters. This innovative theory is also known as "the butterfly effect" and was developed by the American mathematician Edward Lorenz. Lorenz discovered that most meteorological effects are non linear. Therefore this chaos effect would always make it challenging though not impossible to predict weather patterns. The chaos effect explains how small changes of an initial condition parameter may cause an unexpectedly large effect.

In conclusion it seems that minor changes in the solar influx, however caused, might facilitate a full blown ice age. These mechanical parameters are sometimes known as the Milankovitch orbital cycles. However it is possible that reduced insolation may occur via intergalactic dust or bollide encounters. A massive cosmic ray bombardment may occur due to a weakening magnetic field. Cosmic rays are proven to seed clouds. This could lead to a snowblitz.

Our polluted skies might soon reduce the incoming solar radiation in a way that mimics a natural orbital cycle. It has been shown that volcanic aerosols can reduce insolation also. Solar input can be reduced by both aircraft contrails and interstellar space dust from comets. An asteroid impact could lead to a nuclear winter effect. Even burning fossil fuels according to NASA can actually reduce temperatures!

The geologic and ice core paleoclimatic record is irrefutable. The study of deuterium in the Antarctic showed that there were four ice ages during the past 400 thousand years. Each ice age cycle was 100, 000 years in duration and included a short interglacial of around 10 to 12 thousand years. This irrefutable record shows that Earth has been cooling down over the last 65 million years and indicates that our present day interglacial, the Holocene, is due to end shortly.

Chapter 2: Death of the Holocene

Our interglacial, known as the Holocene epoch began when the last ice age thawed. We have now been in our present balmy interglacial for over 11,726 years and time is running out for us. Interglacials seldom last longer than 12, 000 years. The norm for our planet is the ice age or icehouse state as opposed to the green house mode and that is exactly where we are heading. Our beneficent Holocene will come to a dramatic end and much of the Northern Hemisphere will be buried by mile thick ice!

We have already spent around 11,726 years in the Holocene. Hence according to accepted calculations we are about to commence an ice age. Today's warm temperatures and mild conditions are not the norm for chilly Earth during the past several million years. Humans have been exceptionally lucky to have been blessed with such benign and balmy temperatures for the last 11,000 years. Indeed our planet has been in a cooling trend for the past 65 million years and the present warmth is but a tiny climatic blip.

An isotopic temperature record has been derived from drilling deep into the ice cores. This ice core drilling took place in Vostok in Russia in 1998.Isotopic fractions of heavy oxygen and deuterium in snow are temperature dependant. The drilling at Vostok reached a depth of 3623 meters (Petit et al 1999).

Temperature at Vostok,Antarctica. Interglacial periods are marked in green **(Petit 2000***)*

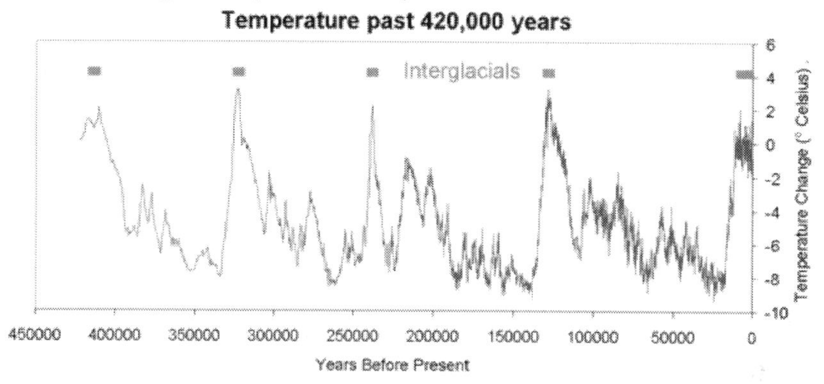

The reader will see that there are deep temperatures troughs and spikes on this graph. The very short plateau top of each spike represents the warmer interglacials. As one can see the top of each spike is rather small in comparison to the plunging troughs of the ice ages. The descending values are much deeper than the plateaus. One can see a precipitous drop going into the ice age. Indeed temperature can be up to 40 times as cold as during the interglacial! You have been warned! However before panic sets in there are other researchers who hope that our expected ice age will only result in temperatures falling by 5 to 10 degrees Celsius globally.

The graph shows that temperature variations are occurring within our interglacial. Despite these fluctuations from colder to warmer periods, our Holocene is considered to be a remarkably stable period of climate for planet Earth. The temperature fluctuations from cold to warm centuries could be considered as mere background noise in comparison to the colossal temperature dip of a looming major ice age.

An important point to make is that it is very unlikely that the global temperature will increase further in the Holocene which is now coming to an end. One can see from the graph that there is a very steep and precipitous descent into the ice age. One can only imagine how such a sudden decline in temperatures might unfold! Another important climatic observation is that our interglacial is the coldest of the five interglacials. This corroborates a theory that in the long term the Earth is on a cooling trajectory. Our Earth is gradually losing heat and this could be very serious indeed.

Within the actual ice age itself the temperature spikes move up and down and these are the Dansgaard-Oeschger events within the ice ages. This book does not cover the Southern Hemisphere in any depth as there is evidence that ice age cycles differ in the two hemispheres in a bipolar fashion. According to CLIMAP data during an ice age when it gets colder in the north it may sometimes get warmer in the south. This divergence of temperatures between the two hemispheres is known as a Dansgaard-Oeschger event.

These events are thought to be caused by the Atlantic Ocean currents and the Gulf Stream which periodically spring to life bringing slightly warmer conditions to the Northern Hemisphere during the ice age. The asynchronous oscillations in ocean currents give rise to temperature differences in both hemispheres. It is becoming apparent that dramatic swings in temperatures are the norm for our planet. In other words the bogeyman "climate change" is perfectly natural and not necessarily something to be feared. The climate will always flip from one state to another and the cause will never be carbon dioxide emissions but rather is driven by the oceans and the water temperatures.

The last plateau table top peak on the above graph is our present day warm interglacial and it is rather wide compared to the other interglacial peaks. This may be an indication that it is about to end. Or perhaps we may dodge the bullet and continue for centuries on this safe plateau of the Holocene interglacial. Certainly we have been enjoying an unprecedented period of warmth which has enabled mankind to flourish and prosper.

Readers will notice that the last interglacial known as the Eemian was warmer and the plateau is therefore higher than the plateau denoting our own interglacial. The Eemian interglacial began 130,000 years ago and ended 115,000 years ago. This interglacial was warmer and lasted for 15,000 years and so was a slightly longer interglacial than usual. Therefore in the historical context we are in a relatively cool interglacial period.

Our present interglacial is actually the *coldest* of the previous recorded interglacials! Our present interglacial has been cooler than the last interglacial. Therefore it is possible that our next ice age may be also cooler than the last ice age, in which case heaven help us! We have now been in our interglacial for the allotted time of an average interglacial, and are at the point of plunging off the precipitous cliff into a deep glacial ravine!

However it is possible, though unlikely, that our interglacial may be longer than the average interglacial and that we may have another four thousand years left to run. This hypothesis is based on research conducted by geologists at Devils Hole in Nevada. Studies at Devil`s Hole caves indicate that we may in fact have been in our interglacial for 17, 000 years already.

Researchers looking at calcite deposits in Devil`s Hole surmise that interglacials might last longer than the accepted 12, 000 years.

However the deep ice core data from the CLIMAP study strongly suggests that interglacials seldom last longer than 12, 000 years. Therefore there is no consensus as to how long our interglacial might last so we need to be vigilant for warning signs that the ice age is arriving.

The warmest part of our Holocene interglacial was five to six thousand years ago. Since then we have been on a cooling trajectory. The middle part of our Holocene interglacial is known as the Holocene Climatic Optimum. The warmer conditions were considered optimal for life to flourish. During our interglacial, civilisation has flourished.

Nomadic hunter -gatherer tribes embraced a settled farming culture and became prosperous. Itinerant nomads settled in one place and so had a steady food supply. They could store the grains in pots and no longer had to keep moving to search for food. These settlements soon became thriving towns and cities. Since the climatic optimum, that was so beneficial to Neolithic farmers, the Earth has been steadily cooling. This indicates that we might be heading towards our next ice age event.

Mother Earth or Gaia actually loves an ice age and oceans thrive during cold conditions. The plant life in the sea ramps up during the cold and this in turn sucks a lot of carbon dioxide out of the air. So as the marine plants thrive the carbon dioxide levels may get lower. The tiny fossils on the sea floor show that more carbon dioxide was stored during the last ice age and this is the reason for this finding. Marine algae play a key part in removing the carbon dioxide from the atmosphere and they thrive during an ice age. So when the ice age arrives it will give the Earth a chance to recover from the damage being inflicted by mankind.

Analysis of the Antarctic Vostok Ice Core discovered that CO2 lags temperature during the onset of glaciations by several thousand years (Petit et al 1999). Please note this is not to be confused with the 800 year time lag of an end ice age thaw. The significance of this observation of a pre-glacial time lag may have been overlooked. Scientists at Cardiff University believe this finding is proof of the role of carbon dioxide in regulating temperatures during ice ages.

The team strove to explain why there is an ice age every 100,000 years (Cardiff 2016). They feel that carbon dioxide plays a pivotal role; but the study is one of correlation and really does not explain what causes the onset of a major ice age. There is certainly a correlation between the slightly lower levels of carbon dioxide and the ice age. However this could simply be an artefact of the biology of ice age marine life and does not incontrovertibly imply causation.

Many present day climate scientists such as those at the Potsdam Climate Institute are relying on our rising carbon dioxide levels to stave off an overdue ice age! However the Vostok ice core data suggests that even as carbon dioxide levels were rising the temperature was falling. The Potsdam Institute for Climate Impact Research in Germany is also pinning its hopes that the modest rise in carbon dioxide will stall our overdue ice age and buy us at least an extra 100, 000 years or more!

Dr. Andrey Ganolpolski from Potsdam is convinced that man can shape the climate and the ice age will be averted. This led to the dramatic announcement by the Potsdam Climate Institute in 2016 that the next ice age has been cancelled! Ganopolski`s team may have got the idea for this study from an article of mine printed in the Oxford Times in 2014 titled "stalling the ice age". Here is a short extract from this article;

“The measured solar levels are now getting so low as to be sufficient to precipitate another ice age. Indeed, before the global warming zeitgeist became popular, scientists were voicing concerns about the imminent ice age. However carbon dioxide levels have risen recently and this may be stalling the expected ice age. This is because carbon dioxide levels might need to be below 280 ppm to facilitate the ice age, and they are presently at about 360 to 400 ppm.”

Since my own *original article* was published in 2014 there have been countless scientific papers published on how humans are stalling the ice age! In 2016 the Potsdam Climate Institute published a study saying that the ice age is cancelled for another 100, 000 years! The lead author Andrey Ganopolski is convinced that the carbon dioxide level is a vital factor in the onset of ice ages. Soon newspapers were excitedly proclaiming that man had managed to avert the next ice age!

However when I wrote my original articles on ice ages, I was actually trying to initiate a debate. I did not believe for one second that our greenhouse emissions would actually stall the ice age. I had merely hoped to draw attention to climate scientists that they had completely overlooked the fact that Earth is due for an ice age. The fixation with global warming had meant that the ice age had been all but forgotten. Soon climate scientists had taken the bait and came up with outlandish proposals that our paltry carbon dioxide level of 400 ppm would stave off the ice age. Unfortunately this will prove to be wishful thinking.

Surely our green house gases will save us from an icy apocalyptic fate? Well let us look at the data we have concerning carbon dioxide and ice. It is known from ice cores studies that levels of carbon dioxide were around 200-280 parts per million during the last ice age. Many scientists are now using this sum to assume that we cannot possibly have our overdue ice age since we now have around 400 parts per million of carbon dioxide. The assumption is that levels of carbon dioxide need to fall below 280ppm before an ice age can begin.

This optimistic assumption may prove incorrect. The reason it may be incorrect is that carbon dioxide emissions may be an artefact of biological activity such as soil microbes. The activity and respiration increases as the climate warms up. It is probable that during warmer periods, life flourished and so more carbon dioxide was emitted. This may be why the carbon dioxide rose during warmer epochs and so the carbon dioxide is an effect rather than the cause of a warmer climate.

The astrophysicist Piers Corbyn also agrees with the time lag theory. He does not think the carbon dioxide by itself caused a rise in temperatures. The data suggests that temperature drops to glacial values well before carbon dioxide begins to fall. This suggests that CO_2 has little influence on ice age inception. A negative feedback loop predominating, will always tip the Earth towards an ice age rather than towards a runaway warming.

As for the melting ice caps which we are constantly seeing on the news this does not in any way negate a coming ice age. One of the reasons that we see huge chunks of ice dramatically falling off into the sea is a process called calving that is perfectly natural. It is also vital to realise that prior to all previous ice ages there has been dramatic melting of ice.

This melting of ice before the onset of an ice age may seem paradoxical and it shows us that climate is never straightforward.

This ice melt water then results in a decreased salinity of oceans which in turn can affect the Atlantic thermohaline circulation leading to decreased land temperatures and an ice age. This is known as a negative feedback loop. It seems that most of the feedbacks are conspiring to promote an ice age in the northern latitudes. There is always an ice age whether ice caps melt or not, so there is no escaping!

The Northern Hemisphere summer now takes place at aphelion. Perihelion and aphelion refer to the closeness of a planet to the Sun. Therefore we receive lower summer insolation at 65 degree north. Cooler summers are one of the conditions needed for an ice age to commence. Glacial inception is thought to start when reduced summer insolation at 65 degrees north allows more ice to survive the summer every year according to axial, orbital type theories. This facilitates the growth of the ice sheets. These are named the Laurentide, Fennoscandian and Siberian ice sheets.

Tens of millions of years ago there was a super continent named Gondwanaland. Gondwana formed when the large land mass known as Pangea split. Gondwana gradually split apart and 50 million years ago the last rift occurred. The separation of present day continents Australia and the Antarctic affected ocean currents making the Earth cooler. A cold frozen Antarctica was formed. It was at about this time that the Transantarctic Mountains were thrust up (Fitzgerald and Gleadow, 1988).

An increased accumulation of snow and ice arrived in Antarctica for much of the Cenozoic Era. Prior to this rift of Gondwanaland, the South Pole was almost tropical! Therefore continental drift and tectonic plate movement has affected our planet`s temperature on a long geologic time frame dating 50 million years. This lengthy time frame helps to put the recent warm decades of the twentieth century into a contextual time frame. A few recent warm years cannot let us escape the fact that our Earth is slowly cooling and that ice ages are becoming longer.

This Holocene Epoch of balmy, clement and stable weather has enabled mankind and our myriad civilisations to flourish. Farming and agriculture started 8000 years BC in the Fertile Crescent and spread to Europe. Livestock were kept in the farming communities of the Holocene. It is thought that the concept of a hierarchy evolved during this time. As people settled into stable communities the opportunity arose to acquire property, wealth and land.

We may not appreciate that climate has been stable during the Holocene when we read of the storms and floods that are occurring worldwide! The Holocene climatic optimum (HCO) was a period of increased warming in the exact midpoint of the interglacial. This period of optimal warmth ended about 6000 years ago and the Earth has been cooling down ever since. So we are slowly gearing up for our expected ice age.

To put this in context let us look at the mini ice age known as the Maunder Minimum. During this period from 1645-1715 the River Thames frequently froze over allowing jolly frost fairs to take place on the icy surface. Yet such frost fair temperature drops are nothing but mere blips on the large timescale of the planet`s temperature graphs. They do not show up at all in comparison with the huge dips showing a precipitous descent into an ice age! The temperature charts show that dramatic oscillations are the norm for planet Earth. By contrast the graph for the interglacial shows a rare levelling out of a high flat temperature plateau.

Scientists are proposing that the Holocene might also be named the Anthropocene because of man`s overriding influence on the Earth and its crust. The term Anthropocene was coined by an ecologist named Eugene Stoermer in the 1980`s. I have often thought how awful it will be for future geologists to find a thin layer of plastic in the sedimentary layer to mark the era when we polluted the Earth with vast volumes of plastic.

Geologists in the far distant future might rename our era the Anthropocene- Plasticene! Perhaps humanity will have died out centuries from now and the geologic layer of plastic and nuclear waste will thankfully never be discovered by future geologists. A major ice age could certainly wipe out a very large number of people. As the crops start to fail in the Canadian breadbaskets of the world there will mass starvation.

The human Anthropocene is leaving a terrible geologic legacy of pollution on our once pristine planet. Perhaps it is a good thing that the Mother of all ice ages may be just around the corner! A really severe ice age could wipe out millions and perhaps enable our long suffering planet to recover. An extinction event such as this could give Earth a chance to survive in a Gaia like fashion.

The Gaia hypothesis was the idea of ecologist James Lovelock. Lovelock felt that the Earth was a self regulating organism striving for balance and harmony in its eco systems (Lovelock 1972). Since human beings are now profoundly disrupting this delicate ecological balance it would make sense that Mother Earth will soon succeed in ridding itself of our troublesome species! Then once more harmony and unpolluted pristine beauty will prevail on our blue planet. In this way Gaia Mother Earth maintains the homeostasis of our ecosystems.

A major ice age would be the perfect means for Gaia to wipe the slate clean to give the planet a chance to recover. Fortunately such an ice age is due. This may please a lot of climate protesters who are worried about the damage being inflicted on our planet. It is rather ironic though, that it will be a cataclysmic global cooling event that will wipe us out, rather than the predicted catastrophic global warming!

The predicted rises in temperature are a result of faulty reasoning and deep misunderstandings about the Earth`s carbon cycle. The geologic processes involving the carbon cycle are very complex. Even the most erudite of scholars and scientists make mistakes with the carbon cycle. This has caused some confusion about the role of carbon dioxide and its correlation with temperature.

It is easy to see how this could have occurred. There are many data sets showing a fairly good correlation with carbon dioxide levels and temperature; but as mentioned previously there are unexplained time lag issues that have been conveniently swept under the carpet. Perhaps this would not matter so much were it not for the fact that fuzzy or loose data is now being used as a basis on which to plan for our future on this planet that is our wonderful home.

The fact that we are due for a major ice age as well as a mini ice age has been completely ignored due to a "consensus" that we are in the throes of a prolonged global warming event. The Holocene has been conveniently forgotten! This is not just any global warming event either. This is a special global warming as it has been allegedly caused by us and hence the term Anthropocene has given rise to the term anthropogenic warming. It seems that like King Canute who thought he could turn back the tides of the sea, humans believe that they are now capable of controlling our climate!

It is now widely believed that fossil fuels have caused a slight rise in temperatures that was recorded during the warm decade of the 1980`s. Warmer years are now thought to have been caused by the Industrial Revolution when coal was king, rather than any natural solar cycle. When coal was king, Great Britain was the richest country in the entire world. However these days coal is definitely a dirty word.

The 1970`s was a decade of growing environmental awareness and soon the attention was diverted away from ice ages to the opposite scenario of a greenhouse effect as concern arose about the possible warming effects of man -made pollution. Sustainability became entangled with the idea of reducing the use of fossil fuels and soon all fears of an overdue ice age were forgotten (Kukla and Matthews 1972).

Before the man made global warming zeitgeist became so popular, there had been countless warnings from respected scientists of an expected return to ice age conditions. During the 1970`s worried scientists were voicing concerns that a return to major glacial conditions should be expected as the Holocene Interglacial had run its expected course. These warnings were soon forgotten.

Not only has our present interglacial run its expected duration, but it has actually exceeded the expected duration of any interglacial. Never before has there been such a lengthy period of climatic stability in our Earth`s climatic history. The odds are now heavily weighted towards the demise of our Holocene. The wildly optimistic hope of climate scientists at the Potsdam Climate Institute that our carbon dioxide emissions will stall this overdue ice age is sadly misguided.

The coldest 1% of the temperatures seen in the past 500 million years on Earth is recorded in the recent Quarternary period. So out of 500 million years, the last one million years are in the coldest range! You do the maths! This cooling process is amplified by the reflective white ice albedo of snow accumulations that progressively helps to cool the Earth. This long term geologic data implies an insidious and ominous cooling of our planet. The worst of the cold weather events may be yet to come as we plummet into a devastating ice age.

Chapter 3: Magnetic Field Weakening: Pole Shift Imminent?

This chapter will explore a possible link between the Earth`s magnetic field and ice ages. The British Geological Survey believes that a magnetic reversal might soon be imminent. The BGS base this on the data of weakening fields such as the South Atlantic Anomaly. The magnetic field has actually been declining for the last 2000 years and most of this decline has taken place in the last 300 years.

Earth`s magnetic field is created by its solid and liquid iron core creating electric currents. The inner core is solid iron but the outer core is a swirling liquid iron that generates our magnetic field. Our magnetic field is driven by the heated currents generated by the iron core which is already a staggering 2.6 billion years old and could even be on its last proverbial legs! As long as Earth`s outer swirling core below the mantle keeps flowing then our magnetic field will be safe.

The entire heated iron core of planet Mars froze out a long time ago causing the planet to lose its atmosphere and become a dead zone. Our iron ore has been cooling by 100 degrees and may have even stopped circulating altogether for all we know! The geodynamo may be about to a stop and we may be living on a dying planet! This idea may not be as far -fetched as it sounds. The magnetic field would continue to be generated for many years after the core had lost its heat. We might not even realise that it had cooled and shut down completely.

If the rotating core packs up, life as we know it will cease to exist on our planet. This is because the magnetic field is needed to protect us from space radiation. A weaker magnetosphere allows more of the intergalactic cosmic radiation to reach us. The cosmic rays are already increasing in a foreboding sign. This cosmic ray influx will be discussed in a later chapter as another factor involved in the initiation of a catastrophic new ice age.

The popular concept of a pole shift refers to a reversal in Earth`s magnetic field from north to south. There is an oft cited doomsday scenario about magnetic North to South reversals and a fear that a pole shift would bring calamity and the Earth would turn upside down! This of course will never happen. Some fear the possibility that the Earth will wobble like a slowing spinning top on its axis. There is an outlandish idea of a wobbling of the Earth on its axis causes seas to slosh around, before the Earth finally settles into its new position! It seems there is no limit at times to the human imagination.

The seas would flood coastal regions potentially wiping out one quarter of the world`s population. It is possible that low lying countries might vanish altogether! Needless to say these exaggerated fears are unfounded. The film Absolute Zero explores the worst case pole shift scenario. The climate of Florida becomes icy like the Arctic as the equator and poles swap climatic positions! Again this is not likely to occur during a magnetic reversal.

However the internal magnetic field does indeed turn upside down during a reversal according to computer simulations. Just how seismically active a pole shift could be is open to debate but there is a possibility that tsunamis and intense storms known as hypercanes might arise. The concept of a hypercane was developed by a scientist from MIT named Kerry Emanuel.

A hypercane is a very intense hurricane that stretches 25 miles high up into the stratosphere. If a meteor hit the oceans it would result in a devastating hypercane according to the model, with supersonic wind speeds of over 500 mile per hour. Extreme conditions in the oceans are needed to form hypercanes. It is thought that a hypercane also helped finish off the dinosaurs 65 million years ago when an extraterrestrial bollide hit the Mexican waters resulting in a mass extinction event.

This meteor impact was preceded by volcanic eruptions causing sulphuric acid rain that destroyed the vegetation and darkened the skies. Not a good time to be a dinosaur! A hypercane would flatten everything in its path and nothing could survive in its wake. A massive underwater volcanic eruption could also cause a hypercane to form as the warm seas evaporate. It is believed that there may be up to three million underwater volcanoes.

A hypercane can funnel up large amounts of dust and water into the icy cold stratosphere which is the freezing upper layer of the atmosphere. Interestingly there has been an increase in the construction of geodesic dome dwellings in America. This geodesic design is held to be able to withstand hurricanes of great intensity and of course the curved roof cannot blow off! Certainly dome homes may be the safest way to build a new house and they are better able to withstand earth quakes and deflect airborne objects too.

The debris from the Jurassic asteroid no doubt caused a nuclear winter sky and blotted out the sunlight killing most of the dinosaurs on Earth. A few tiny mammals survived by burrowing into their shelters underground. This water and dust thrown up by an impact would immediately crystalise into a thin veil of icy reflective diamond dust and facilitate an instant ice age. The astronomer Sir Fred Hoyle was sure that ice ages started with such a diamond dust veil high in the stratosphere. Whether or not a hypercane formed at the KT extinction event it is certain that the debris laden skies caused a cooling of 20 degrees.

The prospect of extreme weather events caused by a pole shift is worrying. Will we have warning of a magnetic pole shift and how can we prepare for such a cataclysmic upheaval. It may be that we are already getting the warnings and yet we are ignoring this imminent peril. According to more optimistic geologists a magnetic pole shift should *not* give rise to drastic continental drift, tsunamis or upheavals of an earth moving tectonic kind. However less sanguine geologists think that seismic activity and flooding will increase with a pole shift. This area is not scientifically settled.

We need our magnetic field to protect us from solar and cosmic radiation. As our magnetic field weakens we become even more vulnerable to major solar storms or coronal mass ejections. One of the more realistic fears is that as the magnetic field weakens during the pole shift it will allow increased amounts of radiation and cosmic rays to penetrate into our atmosphere. NASA scientists have warned that a magnetic pole reversal would leave Earth vulnerable for two hundred years to incoming cosmic ray radiation as the protective magnetic field would dissipate.

Evidence of past magnetic reversals can be found in the solidified volcanic lava sediments on the deep ocean floor and clay pots. Fortunately for geophysicists pottery has been a ubiquitous pastime throughout mankind`s history. Pottery is a useful tool for measuring our ancient magnetic field. Magnetite particles in hot clay line up with the magnetic field at the time the pots are made. As the clay cools the particles are locked into place. Therefore clay pots provide a magnetic time capsule. By examining clay pots scientists have found a very sharp drop in Earth`s magnetic field over the last 300 years.

The Earth`s magnetic field determines how the layers of lava are arranged on the ocean floor. When the magnetic field reverses, the striated layers point in the opposite direction to the other layers. Lava that has been examined is from the Mid Atlantic Rift. The lava from volcanoes provides a record of the past magnetic field. Older lava samples show the microscopic magnets were reversed and pointed due south 730, 000 years ago. When pole shifts have occurred in the past the geologists observe that lava layers point in the opposite direction. Detailed analysis has shown that the last magnetic reversal took only one hundred years to flip.

If this speed of a magnetic reversal were to happen in our 21st century it would wreak havoc on our electrical grid and other technologies. If our electricity supply was cut and colder weather arrived all at the same time we could be in serious trouble. Geologists have established beyond a doubt that magnetic reversals had occurred around every 200, 000 years in our past. Then the pattern changed to the last reversal of 730, 000 years ago which was probably the last reversal (Champion et al 1988).

There has been some consideration given to the idea that a magnetic excursion might have given rise to some previous ice ages. Previous mass extinctions have coincided with sudden magnetic reversals. Past magnetic excursions go by such colourful names as the Gothenburg and Lake Mungo excursion. Evidence for magnetic reversals has been discovered by geologists in the volcanic rock records. During the Permian mass extinction that took place 225 million years ago there was a magnetic reversal. At the same time more than 90 per cent of all life forms on Earth became extinct.

Another magnetic reversal coincided with the famous Cretaceous-Tertiary mass extinction that saw the extinction of the dinosaurs around 65 million years ago. This impact hypothesis is named the Alvarez hypothesis after Luis and Walter Alvarez who discovered the KT boundary. This extinction event is thought to have been caused by a giant meteor impact in the Gulf of Mexico.

A layer of iridium from space has been found in the geologic sedimentary layer. The impact threw up huge clouds of debris. A hypercane may even have formed from the impact. Eventually after the initial firestorm that fried the hapless dinosaurs this meteor debris cooled the Earth and darkened the skies akin to a nuclear winter (Alvarez et al 1980). Certainly not much fun for the inhabitants of the Jurassic park!

Jon Erickson discusses climate and pole shifts in his riveting book called "Lost Creatures of the Earth" which looks at previous mass extinction events. It is thought a magnetic reversal that occurred two million years ago might have initiated the Pleistocene glaciation. The Pleistocene is the geological era that started two and a half million years ago and lasted right up till 11,700 years ago when the ice melted and our present day interglacial commenced. During the Pleistocene epoch, modern humans first appeared on planet Earth.

Three million years ago there was a species of early man known as Australopithicus that evolved from early apes. Australopithicus was followed by Homo habilis who roamed the Earth two million years ago. Homo habilis the handy man ape used stone tools and so was one of first the handy men to inhabit Africa! This early humanoid was later followed by Homo erectus a walking upright human. Then there were the stocky Neanderthals and modern Homo sapiens.

There is a hypothesis that the increase in terrestrial radiation during a pole reversal would have given rise to genetic mutations. These mutations from cosmic rays may have resulted in evolutionary leaps for mankind. Erickson also discusses how the Gothenburg magnetic field excursion coincided with the advance of glaciers and severe cooling. Are magnetic excursions and magnetic reversals potential triggers for ice ages? Might we be heading towards another magnetic reversal event that will simultaneously trigger a new ice age?

NASA scientists alarmingly state that there are signs that we may be heading towards another magnetic reversal or pole shift soon. The magnetic field has been declining for the last 300 years. Research shows that the magnetic field always weakens prior to a pole shift. Computer simulations have now found that loss of magnetic field strength synchronises with the onset of magnetic reversals.

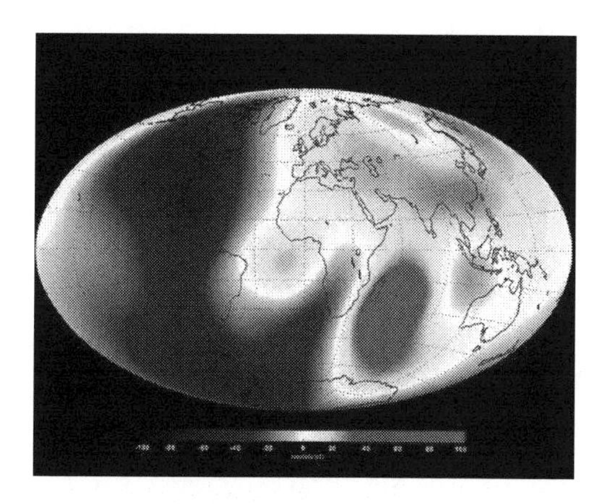

Picture credit: ESA/DTU.

Swarm Satellite picture shows large changes in the magnetic field. dark areas on the left show where the field is weakening.

The mission manager of the Swarm satellites is Rune Floberhagen. He observes that magnetic north is drifting towards Siberia and the field is now weakening at five per cent per decade. This decline is ten times higher than expected and may point to sudden pole shift. This steady loss of field strength is happening at an alarming rate today. The Swarm satellite data can be used to predict earthquakes which may increase as the magnetic field varies.

A magnetic pole shift always follows a decline in field strength according to the work of a Los Alamos geophysicist Gary Glatzmaier. Therefore there can be no doubt that a magnetic reversal will happen since the field is declining rapidly. Glatzmaier has developed computer simulations of magnetic reversals. The simulations show that the magnetic field develops weak patches similar to the South Atlantic Anomaly before it suddenly reverses. A loss of field strength correlates with reversals. We are definitely in the phase of a looming magnetic reversal according to these simulations since there are weak field patches all over the planet (Glatzmaier and Roberts 1995).

As the protective magnetic shield weakens we would see amazing displays of the Aurora Borealis over our skies as the charged particles from the sun reached us. Humans would be bombarded with deadly radiation from space and this could lead to genetic mutations increasing. Our species could become weaker as the radiation takes its inevitable toll on our health. Since the geodynamo is vital to mankind`s existence there have been recent studies to assess the wellbeing of our inner core.

Research into the health of Earth`s dynamic core is presently taking place with a drilling project known as the DEEP project (Determining Earth`s Evolution from Paleomagnetism). The magnetic field needs to be carefully monitored. Evidence from deep sea sediments show that prior to the last magnetic reversal 788, 000 years ago there was a period of great geological instability. We can expect to see increased seismic activity all over the globe and devastating earthquakes as the pole shift approaches.

Another team led by Dr. Andy Biggin from the University of Liverpool has been analysing igneous rocks which reveal the ancient magnetic records. He feels confident that Earth`s rotating outer core will continue to maintain its strength for a very long time. This optimistic assumption is based solely on theoretical models and computer simulations. This is an area open to speculation since no one has been to the centre of the Earth yet!

A professor of geophysics named John Tarduno from the University of Rochester has noticed that our magnetic field has been drastically getting weaker during the last 160 years. He thinks that magnetic flip may be about to happen. He is also concerned about a geomagnetic phenomenon called the South Atlantic Anomaly which is a region of weakening magnetic field that is growing larger.

Tarduno thinks that a region in the tip of South Africa might hold the key to the magnetic field. Tarduno discovered that the magnetic field in the South African region has fluctuated in past years; from 400-450 AD, 700-750 AD, and again from 1225-1550 AD. This South Atlantic Anomaly, therefore, is the most recent display of a recurring phenomenon in Earth's core beneath Africa that then affects the entire globe.

This weakening causes the Van Allen belts to draw closer to Earth which in turn exposes satellites to increased radiation levels. This radiation increase over the South Atlantic Anomaly is damaging to our vital satellites orbiting Earth. A Japanese satellite that was traversing the anomalous magnetic field spun around and blew itself to pieces (Moon 2016).

The Earth is surrounded by the Van Allen belts at a height of around 630,000 miles. These belts trap the cosmic radiation that would otherwise reach us and damage our biological cells. Over the South Atlantic Anomaly the Van Allen belts drop as low as 120 miles above the Earth`s surface. This is low enough to intercept the path of satellites, which need to power down to avoid damage. The Hubble telescope cannot take images while passing over and shuts off.

It has also been reported that laptops have stopped working on the space shuttle as it passed over the South Atlantic Anomaly. The South Atlantic Anomaly covers a growing area over the ocean and has a very weak magnetic field. The weakening field covers Brazil, Paraguay, Uruguay and northern Argentina. There have been strange sightings in South America of celestial lights that have been misconstrued to be UFO`s! Such extraterrestrial UFO sighting events will likely increase as the field continues to weaken. These strange lights in the sky are often caused by tectonic geologic stress as well as magnetic field disturbances.

There have been many frightening encounters with flying saucers along earthquake fault lines! Here in Great Britain there was a fascinating alien encounter that took place in Stoke on Trent which is the site of a geologic fault line. In 1967 a fleet of UFO`s came to inspect the Staffordshire potteries! Terrified locals reported seeing countless spinning objects and strange lights hovering in the skies above them. The Sentinel newspaper excitedly reported that up to seventy sightings had been logged of mysterious globes. The likely explanation is that the fault line was under stress and ET did not come to inspect the world famous potteries!

South America is also home to many UFO sightings and again this is likely due to the magnetic field that is weakening there. Some people even go as far as to claim they have been abducted by aliens! There seems to be no limit to the realms of the human imagination.

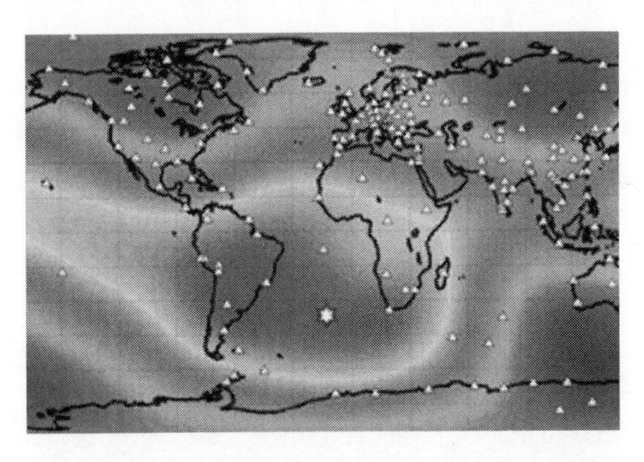

The South Atlantic Anomaly: Credit, Danish National Space Centre

This area is avoided at all costs by satellites and it is now compared to a Bermuda triangle for its spooky effects on aviation. The proximity of the Van Allen belts in this region of the weak magnetic field exposes aircraft to more lightning strikes. A severe storm over this region is thought to have caused Air France flight 447 to crash. An inexperienced co pilot took controls while the pilot had a nap. He was confused by the smell of ozone and flashing lights. The lights were caused by St Elmo`s fire.

This phenomenon is a plasma that is created by a strong electric field. St Elmo`s fire is a form of plasma present during thunderstorms. The icy hail from the thunderstorm caused a malfunction of instruments beneath the plane which has frozen up. The junior co pilot made a mistake and pulled the nose of the plane up. This manoeuvre caused the plane to stall as the plane could not maintain a proper speed. The plane then slowly descended into the sea with its nose still up. This region could be likened to a magnetic pole shift region. Therefore one can surmise that if our magnetic field switched off during a pole shift reversal, the entire globe would be likened to the region of the South Atlantic Anomaly!

This would mean no more Hubble taking pictures of deep space and indeed no more satellites. It is hardly surprising then that the British Geological Survey is taking a keen interest in the area. The BGS has set up camp on a base on South Georgia Island to study the region of weakening magnetic field. One could argue that such research is vital for the future of our planet. As the magnetic field weakens, violent storms and strange weather events become more likely.

The severe weather events that are now becoming more commonplace may very possibly be a result of our weakening magnetic field. These events may be a worrying portent of a catastrophic upheaval known in popular terms as a pole shift. For those readers who may question the veracity of the pole shift concept there is solid evidence of 74 previous pole shifts buried deep in the geological record. So this is not a quasi scientific hypothesis but a very real and terrifying occurrence here on our planet.

A sudden pole shift could cause a mega tsunami on a scale never seen. The waves could even cover the entire land mass of America. It is now accepted that magnetic north has drifted towards Russia away from Greenland. It has moved over 350 miles already in the last decade and continues to accelerate fast. A sudden pole shift could trigger massive eruptions of supervolcanoes such as Yellowstone. The later chapter on volcanoes will explain how this would trigger an ice age.

According to NASA magnetic pole shifts periodically occur on planet Earth. NASA believes that pole shifts has occurred every 200-300 thousand years during the last 20 million years. However this pattern seems to have slowed. The last pole shift, the Brunhes-Matuyama reversal, was late in arriving. If the last pole shift was indeed that of the Brunhes-Matuyama reversal that took place 780,000 years ago then we are overdue a magnetic reversal.

Ice ages used to span cycles of 41 thousand years until something changed around the time of the Brunhes-Matuyama geomagnetic reversal. The ice age cycle then lengthened to one of 100, 000 years in duration. The B-M magnetic reversal took place on Earth 780, 000 thousand years ago. This was the time when early man had just discovered how to make fire! No doubt this discovery helped early man survive the brutal conditions of the ice age.

One might speculate as to why the last magnetic reversal was so long ago. Could this lengthening of time between pole shifts have been caused by the Earth`s dynamo slowing down and cooling? If so could this be the last ever pole shift on planet Earth as our core slowly freezes to death. Could our beautiful planet Earth possibly be dying? The new pattern of 100, 000 year long ice age cycles imply that Earth is slowly getting colder.

Some scientists think that an extremely rapid change in magnetic field takes place rather than a gradual change. It is possible that a comet hitting the Earth can cause a pole shift. This idea was explored in a book by Michael Allaby and James Lovelock called the "Great Extinction". A sudden impact could jolt the Earth and cause the inner swirling core to suddenly reverse direction. Such a fast pace of change would bring disaster to our present day advanced civilisation that is so reliant on satellite technology. Surviving humans would revert to a primitive existence once more.

NASA and other scientific organisations are convinced that the magnetic field is showing a very sharp decline in recent decades. A team led by Paul Renne of the Geochronology Centre based in the University of California Berkeley gave a press release saying that their findings indicate that a magnetic reversal might happen very quickly, within one hundred years or less. Our weakening magnetic field and growing South Atlantic Anomaly could be harbingers of our icy fate. Could this observed weakening of the magnetic field be a sign of an imminent pole shift and a new ice age?

Usually our magnetic field deflects these solar ejections though we see some of their effects when the Aurora Borealis flashes across the polar skies. These major coronal mass ejection (CME) events can blow out power grids. A major CME known as a Carrington Event occurred in 1859. Because there was little technology back then the solar flare caused relatively minor disruption. Earth would be at the mercy of the sun`s coronal mass ejections and cosmic rays.

A similar size solar flare today would bring down all of our satellite technology and we could not even use our credit cards to buy petrol.

We could literally go back to stone -age living conditions. As our magnetic field continues to get weaker we are increasingly vulnerable to a Carrington Event. As our field weakens we will see our skies flashing with the stunning display of the Northern Lights each and every night. The beautiful Aurora Borealis are produced by solar electrically charged particles colliding with oxygen and nitrogen when our magnetosphere is disturbed. The stunning flickering visual displays would be the harbingers of our doom.

We will also see more extreme weather events without our protective magnetosphere. This magnetosphere extends thousands of miles into space and deflects the highly charged cosmic particles from reaching Earth. As the magnetosphere continues to shrink there will be huge tornadoes and giant snowstorms. One such giant snow storm could perhaps initiate our ice age. It could snow for weeks on end or rain giant hail stones for days accruing to a height of a two storey house.

The extreme weather events that we are now seeing may be a direct result of our weakening magnetic field. Rather than attributing these unusual weather events to a 0.03percent of man -made carbon dioxide we should consider how our protective magnetic field is weakening and putting our entire existence at risk. Many people assume that these extreme weather events are being caused by man`s emissions instead of being caused by our declining magnetic field.

While some people are worrying about greenhouse gases the really serious threat we are now facing is our rapidly declining magnetic field. This decline can play havoc with our weather. Scientists speculate that as our field continues to weaken there will be more extreme weather events. More solar radiation and more cosmic rays will lead to violent storms as our protective field diminishes.

To complicate matters further, scientists believe that gases such as carbon dioxide actually increase stratospheric cooling. They say that carbon dioxide in the upper atmosphere facilitates the escape of heat into the cold depths of outer space. So the effect is the exact opposite of the warming effect of carbon dioxide in the lower atmosphere known as the troposphere.

Therefore any increase in carbon dioxide will in the long term add to the cooling of our upper atmosphere. It seems that the odds are stacked against us and we may be heading very soon towards the ice age conditions. Instead of complaining about trace amounts of green house gases in our lower troposphere we should be very thankful for these trace amounts of greenhouse gases. This blanket greenhouse effect in our lower atmosphere is all that prevents our fragile Earth from merging into the freezing cold of outer space. Without this thermal blanket all life would cease on our planet.

Soon we may not even have a protective atmosphere of gases. If our magnetic field disappears altogether then the solar wind can eventually blow our entire atmosphere away into cold space. When this happens all of our atmosphere including the contentious greenhouse gases will be gone forever and life will cease to exist on our beautiful planet. However lest panic descends t is known that Earth once survived without a magnetic field for 3000 years and the atmosphere remained intact.

However our atmosphere was once a lot more dense than it is today. The thermosphere has shrunk by ten per cent in the last 30 years. The thermosphere is cooling and contracting according to observations of satellite drift by NASA. The solar winds are gradually stripping our atmospheric gases into deep space. If all atmospheric gases are lost, the oceans would slowly evaporate and that would spell the end of humanity.

The European Space Agency has a mission called SWARM that is monitoring the magnetic field. As our magnetic field weakens more hydrogen nuclei will enter our atmosphere. These hydrogen nuclei could seed huge snow storms over the cooler latitudes. Such massive super storms could have happened previously during magnetic reversals giving rise to the legends of Noah`s Ark and Atlantis.

A global superstorm of hail and ice could precipitate our overdue ice age. It could hail giant icy stones for weeks on end until there is a twenty feet thick pile of ice on our ground. We should be very afraid indeed. Our modern world depends on satellites that in turn depend on a magnetic field to function. Without satellites our modern world as we know it collapses and humanity might revert to a primitive existence similar to that of the ice age cave-dwelling Neanderthals.

A magnetic reversal today would lead to unimaginable chaos. For example there would be mayhem on the roads as all the traffic lights started to fail. Traffic would be gridlocked in major cities. Then the satellites would fall out of orbit as the exosphere shrinks. Our communications would fail leaving us without television and internet.

In February 2018 a single software malfunction caused chaos to the delivery of chicken to Kentucky fried chicken outlets. A society that is more dependent on technology will collapse faster than more primitive cultures. Technologically advanced cultures will descend into chaos when we lose satellite communications. Nuclear missiles could malfunction perhaps leading to nuclear annihilation.

As the magnetic field first started to weaken the skies would light up with Aurora and flashing lights. As the solar storms continued to batter the unprotected Earth the atmosphere would eventually be stripped away. Eventually the Earth would become a dead planet devoid of life. Without the magnetic field, our magnetosphere supporting the Van Allen belt will collapse.

The deadly radioactive Van Allen belt would then drop down and collapse on us bringing all of the cosmic radiation with it. The Van Allen belt is a zone of energetic charged particles circling the Earth. This belt actually protects us by deflecting cosmic and solar radiation. It is speculated that astronauts crossing the Van Allen zones will receive hefty doses of radiation. This has even caused some conspiracy theorists to doubt that astronauts could have safely crossed the Van Allen belt on the way to the Moon, and by inference they did not even make it to the Moon!

Russian cosmonauts take ginseng to help them survive the radiation perils of space travel. The Apollo lunar mission astronauts noticed flashes of light in their vision and this was most likely cosmic radiation hitting their retinas as they crossed the Van Allen belt. Therefore I think they did indeed travel to the Moon and were bombarded with a few cosmic rays, but returned safely with no lasting effects. Since they did not linger in the deadly zone they would only have received a radiation dose equivalent to a couple of X-rays.

As our magnetosphere collapses, the Van Allen belts are no longer supported and we are at the mercy of threatening space weather! We would be bombarded with a blitzkrieg of cosmic radiation from distant super nova and solar flares. Such a solar flare hit Canada in 1989 when the magnetic field failed to deflect it. This huge solar flare breached the Canadian magnetic field causing widespread power failures as the electricity supplying grid struggled to cope with the massive power surges.

This terrifying scenario has been overlooked in the media frenzy about man -made global warming. A complete magnetic reversal could cause climatic upheaval as our weather shield disappears completely and lead to an ice age as the atmosphere is bombarded with galactic cosmic rays. Look at the barren planet Mars to see what may happen if the magnetic field disappears completely. A pole reversal today would jeopardise plant life on our planet and many crops would fail leading to mass starvation. This is because the intense incoming radiation would kill much of the vegetation on Earth. This lack of vegetation would in turn disrupt the entire food chain. Anarchy would descend as people start to fight over food.

Without a magnetic field, our atmosphere would slowly be stripped away by harmful radiation and the solar wind as happened on the barren planet Mars. Modern life would almost certainly not exist as it does today. Disruption to the protective magnetic field would damage our vital satellite communications sending us literally back to the dark Ages! Ironically we rely on satellites for our weather forecasts!

The upper atmosphere layer is known as the exosphere and it blends into freezing outer space. Space is cold and hostile. You would not want to live in cold outer space! Our protective atmosphere is already thinning. Only the first few miles of our atmosphere can support life and above that pilots require pressurised cabins. The ozonosphere or stratosphere is thinning dramatically and this will facilitate further heat loss that may impel us into a new deadly ice age.

The Ozone Hole

Space is cold, really cold, and we really do not want to be without our protective atmosphere and blanket of greenhouse gases.

Therefore it is alarming to learn of giant holes that are inexorably opening up in our atmosphere. NASA has reported finding giant holes in the ozone layer.

Ozone depletion has been blamed on our use of chlorofluorocarbons in fridges and aerosol sprays. These manmade CFC`s were banned in 1987 in the Montreal Protocol in an attempt to repair the human inflicted damage to our vital ozone layer. The attempt to close the huge ozone hole by banning CFC coolants in fridges appears to have been partially successful. There have been reports that the ozone hole is becoming smaller. In 2017 NASA triumphantly announced that the hole was now at its smallest since the year 1988.

However it was recently discovered that another group of chemicals found in solvents and paint removers can also damage the ozone layer. The new chemical threat is known as dichloromethane. The ozone hole is still the size of two and half large continents such as America. The previous year of 2016 saw the ozone hole grow to an even larger size. Therefore the hole is still gigantic.

Why does this matter? It matters because the ozone layer protects us from deadly space radiation. As our magnetic field continues to weaken the solar winds will punch deadly holes in the protective ozone layer. No amount of environmental legislation to ban CFC`s can prevent this from happening. According to an early study published in Nature magazine, a depletion of the ozone layer will bring about substantial global cooling (Reid et al 1978).

Although the giant ozone hole is mainly situated over the South Pole region there have been reports of huge holes opening up in the North Pole. Ozone holes have even opened up over the U.K. but the Government has hushed it up or at least the event was given little publicity. An ozone measuring station in Lerwick has recorded record low ozone levels of 195 Dobson Units of ozone. This is very low indeed. The same low levels were also recorded by the other ozone measuring station in Cornwall. Despite the Montreal Protocol there are alarming decreases in the ozone layer being observed in the Northern Hemisphere.

In 2017 it was noted that an ozone hole had opened over Great Britain but the observation was not widely publicised. It is possible that the increasing ozone holes are part of the coming ice age cycle and will continue to increase as our magnetic field continues to weaken. Another factor that enables ozone holes to grow is intense cold temperatures in the upper atmosphere. It seems that the upper layers are becoming much colder and this could be a warning sign. We are presently protected from the radiation by our magnetosphere which deflects the charged particles. As our field continues to weaken we will be bombarded with cosmic radiation.

NASA has found that our ionosphere is shrinking at a greater rate than anticipated. When the upper atmosphere is thicker it exerts more drag on orbiting satellites. When it contracts there is less drag. This means that cold outer space is becoming ever closer to planet Earth. Perhaps our shrinking atmosphere is another warning sign of an ice age that is inexorably drawing ever closer.

A team from NASA headed by Mario Ancuna has studied magnetic fields on the planets in our solar system. Mario Ancuna professes that even he is surprised by the precipitous drop in Earth`s magnetic field. There is no explanation for this sudden decline that can be explained by present day theories. Something very strange is going on in Earth`s mysterious core, according to this magnetic field expert.

This scenario of a dying core might even explain why the space scientist Mario Ancuna is so perplexed by the data. Ancuna is baffled by the sudden steep decline in Earth`s magnetic field which has no parallel elsewhere. If the inner core of planet Mars can become moribund then there is no reason to assume that Earth`s dynamic life giving core will last forever. Perhaps this might be the very last ever magnetic reversal if the dynamo of our magnetic field has stopped generating the field. What if our iron core has cooled and slowed to such an extent that it can no longer generate the magnetic field?

This doomsday scenario has already happened on the planet Mars. The red planet has an iron core that froze out long ago leaving a moribund planet with little atmosphere. There is some evidence from ice found that Mars once had water and perhaps basic life forms. However the heated dynamic iron core then froze and stopped generating a magnetic field. This freezing of the geodynamo made Mars a lifeless planet. If the iron core froze in Mars then perhaps this could occur again here on planet Earth. If our core does cool then our planet will go the same way as Mars.

There is no known explanation for the dramatic decline in field strength that we are now recording. The field has never before experienced such a precipitous drop. Could Earth be gearing up for another reversal or even going the same way as Mars? The magnetic field has been decreasing at an alarming rate for the last 300 years. Never before has there been such a steep downward trajectory of the field strength.

However lest panic descends the expert Mario Ancuna thinks it is highly unlikely that our field will disappear. This expert on magnetic fields thinks that the present rate of magnetic field strength decline is far too fast to indicate a dying core. Therefore it must be a sign of an imminent pole reversal.

The fast growing South Atlantic Anomaly is evidence of this serious decline in field strength. We are in dangerous uncharted waters. In 2016 it was reported by the TATA Institute in India that a large crack had appeared in our protective magnetic shield. The GRAPES- 3 Muon telescope recorded that the crack allowed a large burst of cosmic rays to reach Earth. The magnetosphere became very compressed and left us vulnerable to the perils of space weather storms.

The magnetic field not only affects our upper ionosphere but it also influences the wind patterns in the lower atmosphere layer called the troposphere. The troposphere is where most of our weather events take place. It is speculated that these cracks in the field will give rise to superstorms. Such a superstorm was seen in Australia recently in the year 2017 when a category 5-6 cyclone battered Queensland. Cyclone Debbie battered the town at a record 190 miles per hour!

When the main field collapses the levels of incoming space radiation will skyrocket as the field is no longer able to deflect the space weather. Every single night the skies would light up with spectacular displays of the Aurora Borealis. The Northern Lights would no longer be the exclusive preserve of northern latitudes. People would be able to look up at the night sky and see the shimmering borealis! This would be a beautiful sight though filled with peril. There is no dodging this bullet and we need to prepare for this cataclysmic upheaval. Our long awaited magnetic reversal is now due and the ice age awaits us.

Chapter 4: Our Sun is Powering Down

At the beginning of March 2018 Great Britain experienced unexpected snowfalls due to a change in our usual wind directions. The infamous "Beast from the East" winds from Russia brought record snowfalls in March. This change in wind direction from warm westerly to cold easterly is thought to be affected by the sun. Professor Mike Lockwood is convinced that a low sunspot count can induce more cold easterly wind flows. In 2017 Mike Lockwood at the Rutherford Appleton Laboratory in Oxfordshire said that the sunspot activity is declining at its fastest rate for *ten thousand years*!

Could this be the start of our new ice age? Professor Mike Lockwood of Reading University is concerned about the declining strength of our Sun and its heliosphere. The heliosphere is a protective bubble that deflects nasty space weather away from Earth. Space weather is loaded with deadly cosmic radiation and energetic gamma rays. Lockwood believes that the heliospheric bubble that is protecting both planet Earth and our solar system will have substantially shrunk by mid century. Then we will be at the mercy of whatever the cosmos decides to throw at us.

The mid century year 2050 is also predicted to be the culmination for three decades of a solar sunspot minimum that may trigger a mini ice age. Professor Mike Lockwood is a British atmospheric physicist who has carried out much of his research at the Rutherford Appleton Laboratory in Oxfordshire. Lockwood has observed a sustained decline in the sun`s activity. Indeed he believes that the sun's activity is in free fall. Solar activity is declining faster than at any time in the past 10 thousand years according to Mike Lockwood, professor of space environmental physics at Reading University.

British scientist Mike Lockwood from the Rutherford Appleton Laboratory believes that solar activity affects our jet stream weather patterns. The jet stream is a band of incredibly fast moving westerly winds high up in the atmosphere which circle around the pole. It can feature winds of up to 200 knots (230 mph) and these winds tend to guide wet and windy weather systems which come via the Atlantic.

The jet stream moves around and its position can have a big impact on weather here in the UK. If the jet is over the UK or just to the south, we tend to get a lot of wet and windy conditions as it brings weather systems straight to us. The floods in the British Isles of 2014/15 were caused by a powerful jet stream. Cold easterly winds from Siberia are influenced by low sunspot activity. In February 2018 an excited press announced that the Siberian "beast from the east" was about to bring freezing blizzards of sleet and snow to Great Britain!

The Dalton minimum was a cool climate sunspot cycle from 1790 to 1830 discovered by the astronomer Jon Dalton and his wife Annie Dalton. Another low sunspot cycle was known as the Maunder Minimum that took place from 1645 to 1715. Let us not underestimate the biting cold of the Maunder Minimum which was bitterly hostile to life. The sunspot minimum happened at the same time that weather in Northern America and Europe became unusually icy and cold. The Maunder Minimum occurred in the 17th century and lasted around 70 years.

During this time Londoners enjoyed skating upon the River Thames. The citizens that lived in 17th century Europe survived these cold winters, without the heating and insulation technology that we are fortunate to have today. If the next solar activity minimum does affect the weather on Earth, perhaps it will not be so unpleasant for Londoners. However if another little ice age were to morph into a great ice age things might not look so rosy.

The Maunder Minimum was studied by the astronomer John A. Eddy who was a solar astronomer. Eddy demonstrated that our sun is not a constant star with a regular cycle but rather one that has periods of anomaly and variation. In 1976, the astronomer Dr. Eddy published an article in the journal *Science* in which he confirmed the speculative observations of 19th-century astronomers.

Eddy noted that from 1645 to 1715, the surface of the Sun was inordinately calm. The sun`s magnetic storms which are indicated by sunspots were peculiarly absent. “I have re-examined the contemporary reports and new evidence which has come to light since Maunder’s time and conclude that this 70-year period was indeed a time when solar activity all but stopped,” Eddy wrote.

Is it really possible that a lack of sunspots could have caused such a drop in the temperature in Great Britain? A sunspot is a cooler area on the surface of the sun in the photosphere that appears dark.

There appears to be a correlation between the number of sunspots and solar radiation. Sunspots have a hypothesised effect on our climate and sunspots also provide useful information for space weather forecasts that enable the safety of satellites to be monitored.

The future predicted activity of the Sun in the coming decades has been likened to the Maunder Minimum. There was also another cooling period after the Medieval Warm period during the five centuries long period known as the Dark Ages. So it appears that there have been quite a few cool periods in the past. It looks as though we may be heading towards another such cool climate event if the solar physicists are correct in their predictions about declining solar activity.

In the not too distant future the River Thames might once more freeze over. In days gone by the River Thames was the scene of frost fairs and market stalls were set up on the icy waters. This could be a reality once more for the citizens of London. Perhaps an ice rink or two may also appear on the River Thames. Recent structural and engineering changes to the London bridges perhaps make it less likely that the River Thames will freeze over since it now flows faster. So maybe we will not live to enjoy the ironic spectacle of our elected politicians gaily skating on the River Thames!

With all of the talk about man-made global warming one would think that humans are in control of the climate and that our Sun has a mere peripheral role. Indeed our star that is powered by thermonuclear fusion has been relegated to a bystander role in the warming of our planet. This is plainly nonsense.

An interesting Oxford lecture on climate took place in 2004. It was given by the author Mark Lynas who was promoting his book "High Tide" sensationalising the perils of rising seas caused by global warming. He later wrote an even more sensational version of this apocalyptic theme called "Six Degrees; our future on a hotter planet". It was a very entertaining talk and this talented author and adventurer knows how to captivate his audience.

I was full of admiration for the fact that he had left Oxford with his proverbial knapsack and gone to explore the world in search of melting glaciers and other evidence of global warming. I once sat next to the budding author at a party in Oxford when he suddenly announced to an invisible spectre looming before him that he was going to write a bestseller. He had the mystical otherworldly look of a sublime visionary who had been visited by a bedazzling muse of creative inspiration.

He certainly did not appear to be present in the real mundane world when he decided to become a bestselling writer! I later bought his book "High Tide" and would recommend it as a riveting read. I was therefore interested to attend the Oxford lecture in 2004. However I soon felt that the role of our sun had been completely overlooked during the presentation and some members of the audience also thought so.

I recall a member of the audience politely raising his hand and tentatively enquiring about the role of the sun in our climate? Was the author really sure that our carbon dioxide emissions were the main driver of the climate, he nervously asked. Never a person to be fazed by an awkward question the confident author brushed the topic aside like an annoying gnat buzzing around his person. The sun it seems is not that important at all as a driver of climate, he opined. Tides will rise and flat Holland`s colourful tulip farms would be washed away under a tsunami caused by global warming. It seems it was not the sun keeping us warm after all but the warming greenhouse emissions from our coal fires.

Not everyone was entirely happy with the downplaying of the sun`s role in climatic variables , but on the whole the audience left the Oxford talk feeling suitably entertained and satisfied with their outing. The book promotional talk had presented us with an agreeable frisson of dangerous calamity as well as the prospect of a balmy Bermuda like climate replacing the boring grey skies of Great Britain. Plus there had been nice slides to look at and very comfy plush leather chairs to lounge on in the Turl Street club.

However most of the audience left the talk endowed with a healthy degree of cynicism. That is not to say that I do not agree that it might be dreadful for our entire planet to warm by six degrees. However the opposite scenario of a global cooling by minus six degrees would be equally horrific and bring a major ice age to the Northern Hemisphere. Indeed we are only five degrees away from entering an ice age right now and there are some worrying signs that our sun may be going into a period of hibernation.

In 1990 NASA and the European Space Agency joined forces to send a satellite to orbit the sun. This robot satellite was called Ulysses and provided some unique data about the sun as it made three complete orbits before it was decommissioned in 2009. The robotic probe was also collecting data about the heliosphere which is the entire region of our solar system that is dominated by the sun`s energy.

The solar wind extends right to the far edge of our solar system beyond Pluto where it eventually pauses at it encounters interstellar space. At this junction it becomes known as the heliopause. The heliopause arises when the solar wind has become too weak to counteract the interstellar medium. There is a large increase of cosmic ray influx at the heliopause due to the weakness of the solar wind. Lest my readers start to yawn I should explain that all of this is very important for our climate here on Earth. A very interesting finding was that the solar wind was measured as becoming progressively weaker during the twenty year mission.

The magnetosphere is crucial to deflect various nasties coming at us from deep space. These nasties include cosmic rays that are highly radioactive. These rays are mutagenic particles that emanate from the Galaxy. Cosmic rays are present at a high altitude in Earth`s upper atmosphere. Passengers in high altitude aircraft may be at risk if they are frequent fliers or are pregnant. However a healthy buoyant solar wind is able to deflect most of the harmful galactic cosmic rays away from Earth`s upper atmosphere.

In 2008 it was reported that the solar wind was at its weakest strength since recordings had begun in 1969. The solar wind had declined in strength by 20 percent. One of the principal investigators for Ulysses was Dave McComas who is based in San Antonio, Texas. McComas was overseeing a solar wind sensor aboard Ulysses called SWOOPS. He was shocked to record such a large drop in the solar wind strength.

As the mission was uncovering a lot of interesting data you would have thought that NASA would continue to fund this worthy mission. Sadly this was not the case and the funding was withdrawn. Perhaps the funding was diverted to study carbon dioxide levels back on Earth! This decision is extremely unfortunate as if the solar levels continue on a downward trajectory it may have important repercussion here on Earth.

The repercussions on our climate of a quiescent sun may turn out to be far worse than any hypothetical anthropogenic warming. This winding down of our sun`s strength may be the warning sign that we are soon to enter into a new ice age. This conjecture is given further weight by the fact that cosmic rays are thought to influence cloud formation on planet Earth. A weaker solar wind allows more of these galactic cosmic rays to penetrate our atmosphere.

Thus when the cosmic ray flux increases due to a weakening solar wind there is the potential for more clouds to form. The whiteness of the clouds in turn may provide a greater albedo effect as the sun`s rays bounce back into cold space. Thus there is feedback loop whereby a weakening solar wind may indirectly lead to a lowering of Earth`s temperatures. This idea was fully explored by Nigel Calder and Henrik Svensmark in their excellent book "Chilling Stars, a new theory of Climate Change". The former editor of New Scientist, Nigel

Calder, makes a strong case in this book for the sun being the main driver of climatic variables as opposed to carbon dioxide.

The sun`s heliosphere is created by the solar wind and it is now drastically is shrinking. The heliosphere reaches to the very outer reach of our solar system. The heliosphere is protecting our solar system from the high intensity cosmic radiation from outer space. This heliosphere is generated by electrically charged particles travelling at speeds of a million miles an hour from the sun. At the far reaches of our solar system a shock wave forms that deflects the interstellar radiation.

The heliospheric solar wind has been measured as being 20 per cent less dense by the Ulysses mission. Since the solar wind is no longer inflating the entire solar system heliosphere there is much less shielding from cosmic rays. This is very worrying as there has been a lot of research by Henrik Svensmark showing that cosmic rays may lower global temperatures. The cooling influence of increasing cosmic rays on our climate will be discussed in a later chapter.

The heliosphere is vital to life on Earth and without it life would be impossible here. Yet despite its importance to our existence a decision was made to withdraw funding for the Ulysses shrinking heliosphere study! The Ulysses Mission funded by NASA and ESA shows the solar wind is cooler and less dense than one might expect.

Ulysses also discovered that dust coming into the Solar System from deep space was 30 times more abundant than previously expected. This space dust could also exacerbate the trend for a new ice age as the ultra fine micron particle dust veil reflects light back into cold space. The unpublished data from the abandoned Ulysses project found that the number of cosmic ray particles jumped by 20 per cent in our upper atmosphere. This is a huge increase of a deadly particle. The Ulysses mission also found that the sun`s underlying magnetic field has been weakening by around 30 percent since the 1960`s.

In 2007–2008 *Ulysses* determined that the magnetic field emanating from the Sun's poles is much weaker than previously observed (NASA briefing 2008). A weakening of the magnetic field shrinks the heliosphere emanating from the sun enabling more cosmic rays to enter our atmosphere. Ulysses discovered that the Sun's magnetic field interacts with the Solar System in a more complex fashion than previously assumed.

Since this worthwhile mission was rashly halted in 2009 we can only guess as to what is now happening to the solar wind. This shrinking of our the heliosphere would mean that any future voyager missions would be able to reach the region outside our solar system that is known as interstellar space in a shorter time. In other words the freezing life threatening cold region of intergalactic space is drawing ever nearer to our planet that is presently thriving happily in the Goldilocks zone!

The magnetic fields of both the Sun and the Earth are thought to be generated by similar "dynamo" processes that involve rotating conducting fluid. The field is generated by molten iron in the case of the Earth and hot ionized gases in the Sun. The Wilcox Solar Observatory has observed the sun`s magnetic field since 1975. The data from the observatory show that the sun`s two hemispheres are strangely out of synch at times. Intriguingly changes in the sun`s magnetic field affect lightning strikes here on Earth!

The sun`s magnetic field can warp Earth`s magnetic field. This in turn leads to increases in cosmic rays that precipitate thunderstorms and lightning. The seeding of thunder clouds by cosmic rays creates lightning bolts of electrical energy. The high energy particles called galactic cosmic rays provide the link that lets the current flow into a lightning bolt. The sun`s magnetic field impacts Earth`s magnetic field in a way that triggers electrical thunderstorms. A single cosmic ray is a proton that has energies a million times higher than anything produced in the Large Hadron Collider at Cern in Switzerland.

The galactic cosmic rays provide a wire-like conducting channel of ionisation that charges lightning bolts. Lightning strikes here on earth are triggered by energetic particles brought via the solar wind. Astrophysicists have observed that cosmic rays can affect weather on Earth. The link between the Sun`s activity and thunderstorms can provide a useful tool for predicting our weather. So when there is a thunderstorm blame the sun for a magnetic field disturbance! The frequency of lightning strikes follows regular patterns that match the rotation of the Sun`s magnetic field.

David Hathaway of NASA Marshall Space Flight Centre and Matthew Penn of the National Solar Observatory have also found a decline in solar activity. The team presented their unexpected findings at the July 2013 meeting of the American Society Solar Physics Division. Hathaway observed a weakening of the solar cycle 24. This was a surprising finding as in 2009 Hathaway thought that the future solar cycle 24 would remain strong and buoyant.

The Russian scientists Zharkova and Popova have recently replicated this seminal 2013 American study of diminishing sunspots. The sun has recently put on its weakest performance for one hundred years according to Hathaway. This rapid powering down is extremely worrying. Could this data be warning us of a major glaciation event? **Cycle 24 is the weakest for 100 years**, according to D. Hathaway of NASA. Matthew Penn came up with a catastrophic hypothesis to account for the data. Penn speculates that the sunspot cycle might die completely. His predictions are based on the weakening magnetic field within the sunspots. David Hathaway based his measurements of the Sun's polar field and the *meridional flow* which is the flow of magnetic flux from the Sun's equator to the poles.

A stronger flow would help strengthen weak fields, but meridional flows have been completely absent in cycle 24. As far back as 2008 it was observed that there were 266 days without any sunspots! Both Hathaway and Penn expect the coming solar cycles to be the weakest recorded. Worryingly David Hathaway thinks that the last Maunder Minimum was a catastrophic event causing mass starvation.

Matthew Penn and William Livingston predict sunspots may vanish when the magnetic field strength falls below 1500 gauss.

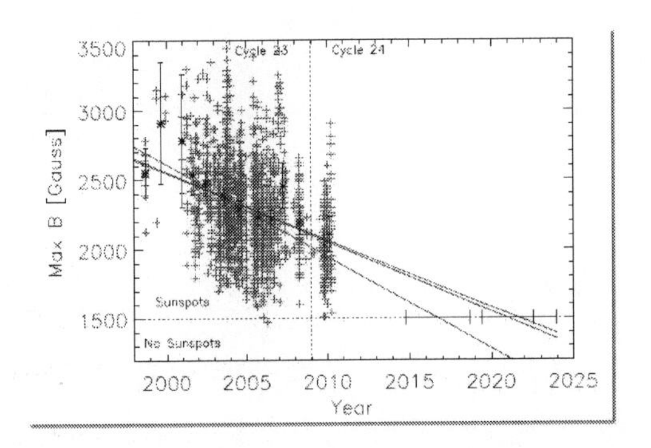

Penn and Livingston 2010

The astrophysicists made a linear extrapolation of the magnetic field trends. This suggested that the threshold of 1500 Gauss would be reached by the year 2017 (Penn and Livingston 2010). As you can see in the second diagram the projected solar cycle 25 shows a collapse of sunspots and perhaps will mark the start of a major ice age!

While some scientists are debating whether we will see a return of the mini ice age weather patterns, they may be missing the huge elephant which is the next major ice age. Lockwood and his colleagues are wondering whether the decline will become the first "grand solar minimum" for four centuries.

During a grand minimum the normal 11-year solar cycle is suppressed and the sun has virtually no sunspots for several decades. There should have been a peak in the number of sunspots, but it didn't happen. Professor Lockwood thinks there is a 25 per cent probability of Maunder Minimum conditions when there were no recorded sunspots for 70 years!

The National solar Observatory in Tucson Arizona has recorded a drop in magnetic sunspot activity since 1995. William Livingston and his colleague Matthew Penn have recorded a slide in sunspot magnetic fields for which there is no explanation. The Solar and Heliospheric Observatory, SOHO, mission has recorded a drop in the actual output of the sun by 0.015 per cent.

This fall in solar irradiance was measured by the Virgo instrument. This may sound small but actually depicts a huge fall in solar output. A fall in output actually suggests that the sun may be shrinking! The amount of heat the sun gifts us here on planet Earth depends on the second power of the solar diameter.

Therefore a minuscule change in diameter could throw the Earth into an ice age. This concept of a shrinking sun was explored in the film *Sunshine* where a team of scientists embark on a perilous journey to the sun in the year 2057. Their aim is to reignite the dying sun with a nuclear bomb!

Apart from the data provided by the Ulysses satellite there has been more recent data coming in from solar physicists regarding the sun`s activity. A recent study by Professor Valentina Zharkova at the Wilcox solar observatory observed that our sun is reverting to a less active sunspot phase comparable to the low sunspot number observed during the Maunder and Dalton Minimum.

Zharkova thinks that the next sunspot minimum will last for 30 years. The cautious astronomer predicts that the climate might become cooler in Great Britain. “The conditions during this next predicted minimum will be chilly. “It will be cold, but it will not be this ice age when everything is freezing like in the Hollywood Day After Tomorrow films,” Zharkova says!

There are some scientists who believe that something far worse may be about to happen! Zharkova`s team monitored the Sun's background magnetic field which causes solar features like sunspots. After analysing the solar data Zharkova's team noticed something that no one had observed before. It was observed that the Sun produces its magnetic waves in pairs. Previously researchers had thought that there was only a single source of magnetic waves in the Sun, but the new groundbreaking evidence suggested two sources.

The team used these observations to predict how the Sun's magnetic field would change in the future. This is how the team predicted the new Maunder like solar conditions that will arrive in the year 2030. The conditions will thereafter continue to get worse. During the solar minimum, the intensity of solar radiation will be reduced dramatically. So we will have less heat coming into the atmosphere, which will in turn reduce the global temperature. Will this be the start of our overdue ice age?

Zharkova and her colleagues use a technique called principal component analysis of the magnetic field. The researchers analysed magnetograms for three cycles of solar activity by applying principal component analysis. This reduces the data dimensionality and noise to identify waves with the largest contribution. This method can be compared with the decomposition of white light on the rainbow prism detecting waves of different frequencies. This method reveals that the magnetic waves in the Sun are generated in pairs (Zharkova et al, 2012, MNRAS).

They observations were recorded from the Wilcox Solar Observatory in California. Zharkova`s team examined three solar cycles of magnetic field activity, covering the period from 1976-2008. They compared their predictions of temperature to average sunspot numbers which are a strong marker of solar activity. The predictions and observations closely matched. Looking ahead to the next solar cycles, the model predicts that the pair of waves increasingly offset during Cycle 25. The Russian scientist calculated the summary curve linked to the variations of sunspot number.

This solar cycle 25 peaks in the year 2022. During Cycle 26, which covers the decade from 2030-2040, the two waves will become exactly out of synch. This phase is predicted to cause a significant reduction in solar activity. This solar cycle is the start of our mini ice age cycle. When there is full phase separation, we have the very cold conditions last seen during the Maunder minimum, 370 years ago.

Another astronomer, Dr. Helen Popova, agrees with the sunspot theory. She also spoke at the North Wales astronomy conference in Llandudno in 2015. She has developed a similar mathematical model of the Sun`s magnetic activity. The model also indicates a reduction in sunspots during the future solar sunspot minimum that could lead to a cooling of the Earth`s atmosphere. Popova agrees that the principal component pair is responsible for the variations of a dipole field of the Sun.

The magnetic waves are assumed to originate in two different layers in the sun (Popova et al, 2013). Both scientists correctly predicted that sunspot magnetic activity would decrease in the present day solar cycle 24. This forecast of low sunspot activity has proven to be accurate in the year 2018. Our present day cycle has been the quietest for over a century. This quiescence could be an ominous sign of things to come (Zharkova et al 2015).

The Royal Astronomical Society (RAS) concurs with the Russian solar physicists and predicted that solar activity will fall by as much as 60% in the 2030`s to 'mini ice age' levels. A new model of the Sun's solar cycle is producing unprecedentedly accurate predictions of irregularities within the Sun's 11-year heartbeat. The model draws on dynamo effects in two layers of the Sun, one close to the surface and one deep within its convection zone. Solar cycle 25 will take place in the years 2019 – 2030 and solar cycle 26 takes place in the years 2030-2040. The dynamo model of the sun was originally first observed by Mausumo Dikpati` s team in 2004.

The RAS model of the sun confirms that it is driven by a double dynamo (July 9, 2015 Royal Astronomical Society). The RAS agrees that the waves will become fully separated in opposite hemispheres in cycle 26 and thus have little chance of interacting and producing sunspots. This will lead to a sharp decline in solar activity in years 2030-2040 comparable with the conditions existed previously during the Maunder minimum in the XVII century. During this freezing and impoverished time there were only about 50-70 sunspots observed instead of the usual 40-50 thousand.

This is a massive drop in numbers that will lead to a reduction of the solar irradiance by $3W/m^2$ according to solar physicists. This reduction in solar input of 3 watts per meter squared is actually rather large when one considers that the IPCC say that carbon dioxide will manage to somehow heat Earth by 1.4 watts per meter squared! If these figures are correct then mankind could be in for a nasty shock as the temperatures plummet in the Northern Hemisphere.

This reduction in sunspots could lead to a new mini ice age. This predicted solar irradiance reduction might also be exacerbated by the increasing cosmic radiation and a global dimming that is caused by our polluted skies and volcanoes. Together the many contributing factors could amplify the trend for cooling in a deadly cooling feedback loop. Such a large reduction in solar radiance far outstrips any insulation reduction from a Milankovitch cycle. Therefore this prediction is rather worrying to say the least!

If the theories about the impact of solar activity on the climate are correct, then this sunspot minimum will lead to a significant cooling, similar to the one that occurred during the Maunder Minimum. To reiterate, the Maunder Minimum coincided with the coldest phase of the 500 year period known as the "Little Ice Age" which began in 1315. This cold period was preceded by a medieval warm period. This warmth lasted from 900 to 1315 A.D and resulted in a flourishing prosperous era of enormous productivity and great Cathedral building. Then in 1315 the balmy climate suddenly disappeared in Northern Europe and crops failed after snow and torrential rains weakened them.

This sudden change in the climate was followed by a great famine in the years 1315-1317. The Little Ice Age continued for 500 years until it ended around 1850.The foul and thunderstorm ridden weather was sometimes blamed on witches who were accused of brewing up diabolical storms! Temperatures dropped by ten degrees and crops had become waterlogged. Every devastating hail storm was blamed on sorcery and witchcraft, since the science of meteorology was unknown. "In 1445, was a very strong hail and wind, as never seen before, did great damage; many women, said to have made the hail were burned as witches according to the law" (Behringer 1999).

The Astronomer, William Herschel, tracked the price of wheat and bread and how these varied with sunspot activity. The astronomer found that fewer sunspots resulted in lower yields of wheat. During the Maunder Minimum there was often heavy rainfall during the summer months that made crops water logged. Millions of people died from starvation during the coldest years which were from 1645 to 1720. Think of our very wet summer in 2012 in Great Britain and this might give an idea of how a mini ice age summer would have ruined crop yields.

These days there is a tendency to equate floods and stormy weather with a manmade warming. Yet the evidence points to this pattern of very wet stormy weather as being prevalent during the cold Little Ice Age. During this period there were very cold wet winters in Europe and North America. Waterlogged crops resulted in famine. The 500 year long little ice age culminated in the cold Maunder Minimum when the Thames and Danube Rivers froze. The Moscow River was covered by ice and Greenland covered in glaciers in the 17th century. During this period, there were only about 50 sunspots instead of the usual 40-50 thousand sunspots which is a huge drop in numbers.

The Sun has its own magnetic field. The amplitude and spatial configuration of the magnetic field vary with time. This results in changes of intensity of plasma flows coming from the Sun, and the number of sunspots on the Sun's surface. The number of sunspots on the Sun's surface has a cyclic structure that varies every 11 years. This cycle is indicated by the analysis of carbon-14 and beryllium-10, glacier isotopes. There are several solar cycles with different periods and properties. As well as the 11 year sunspot cycle there is thought to be a 90 year cycle and a 206 year cycle.

It seems that our Sun is far from stable and highly prone to variation. NASA and the University of Arizona measured magnetic-field measurements 120,000 miles beneath the sun's surface. Their findings also suggest that solar Cycle 25, due in 2020, will be a great deal weaker still. Solar cycle 26 which is arriving in the year 2030 is expected to be the smallest in over 300 years. As for solar cycle 27 one can only speculate how cold it might be if the sun`s dynamo does not recover!

It is interesting to also speculate as to the cause of the simultaneous decline in our Sun`s magnetic field with the Earth`s magnetic field. Could the two declines in field strength be somehow connected and be a terrible warning of our impending ice age? This decline could coincide with the dawn of a new mini ice age or worse.

We could experience a dramatically fast transition to an ice age if the pollen and plant records are correct. Pollen records show how plants changed from warmth loving species to cold weather varieties in a very short time frame. A variable output star like the Sun could literally become dim. We could wake up one day in Great Britain to find our sun looks eerily pale and wan at the dawn of a new ice age.

The solar cycle that we really need to look out for is solar cycle 26. This cycle begins in the year 2030. This could be the year we notice a return to very severe cold weather conditions last seen in the 17th century when the River Thames would freeze. Imagine how surprised our politicians would be when they look out of the Houses of Parliament onto a frozen river. Especially imagine the reaction of the politicians who were involved in climate legislation to save carbon emissions to prevent a rise in temperatures! Instead of the predicted warming, there will be a deep freeze in the middle of this century!

Since solar cycle 26 is being heralded as the harbinger of a dramatic drop in temperature it may get even worse by the year 2050. Some of us may live to see the dawn of a huge glacial inception that heralds the dawn of our major overdue ice age. All fears of a global warming will be long forgotten as we struggle to adapt to a massive ice age. A major ice age is far worse than a mini ice age lasting a few decades, as it usually lasts for at least 90, 000 years!

Pensioners may say at their coffee morning socials in the year 2050 "oh I remember in those good old days when we were going to have a nice global warming and grapes growing in our country!" "Oh yes I do remember those good old happy days when the Greens were protesting about global warming and throwing cream pies at Jeremy

Clarkson! Now that was a lark! We used to surf the web for hours but now there is no power for most of the day since the coal power stations were closed to prevent global warming and the wind turbines have frozen over".

"Ah yes I really loved those good old days when we had our lights on, lots of electricity all day and hopes for a nice warm country. Didn't that James Lovelock chap say we would be boiling hot by now?" "Well Freya, you know how these climate experts always get things wrong. Anyway I`m off home now; would you like a ride on my new snowmobile?"

Pensioners may also start to die in increasing numbers as the cold winds bite. The elderly are more vulnerable to freezing temperatures and already the winter mortality rate is unacceptably high in Great Britain. Life in a freezing Great Britain may come as a nasty shock. Pensioners might be wise to consider migrating to warmer Southern climes such as the Spanish Costas, though they too might succumb to the ice age.

Let us hope we will not end up suddenly frozen like the woolly mammoths found with food still in their mouths as they suddenly froze to death. No one knows for sure why the mammoths and mastodons died out suddenly overnight while eating their dinner. Millions of mammoths have been found buried in permafrost and the indications are they possibly could have frozen to death instantly as many were still standing.

Many of the deep frozen mammoths had broken bones and this could have been caused by gigantic hailstones, during an ice storm that killed them instantly. It is unlikely that the mammoths were eaten by hunter gatherers as their remains were found in very unnatural positions. The suspicion is that they all died instantly probably in a giant ice storm with hail the size of golf balls.

We have to consider the awful possibility that a major climatic upheaval could result in a mass extinction event. Could an ice age come on suddenly within months as has been suggested? Could this coming cycle of solar sunspot quiescence herald the dawn of a major ice age unlike anything that mankind has ever seen? Cycle 26 may be the trigger for the inception of our overdue great ice age that arrives at the end of every interglacial.

This deadly ice age is now due to begin according to data from the geologic record. Since we are now overdue a major ice age, this solar sunspot cycle may herald the dawn of a major ice age that just happens to commence along with the expected mini ice age. A major ice age will be up to 30 times colder than a mini ice age event such as that of the Maunder Minimum!

Our Sun may not be a normal star that shines and irradiates in a homogeneous unchanging fashion. The Sun instead may vary in its luminosity. It might fluctuate in a similar way to variable stars known as Cepheids. These Cepheid stars fluctuate periodically in a way that depends on mass and luminosity. Our Sun has many phases and cycles and is chronically unstable. The Sun is almost human like in waking up and becoming fully alert and then falling asleep and becoming dormant. Maybe the idea of the Sun becoming weak and faint is not so far -fetched after all.

Some scientists take this concept of a variable star to a whole new level. They believe that the Sun is affected by primer fields from interstellar space. When the primer fields become inactive the Earth`s climate can become 40 times colder than it is at present! Scientist Rolf Witzsche is convinced that the Sun`s primer fields are showing signs of a rapid decline.This will plummet us into major ice age conditions by the year 2050. This will be far colder than a mini ice age.

Not everyone agrees with the primer field theory of plasma cosmology. However the declining solar output predicted does seem to coincide with the accepted Ulysses data on measured declining solar winds. The retired scientist Rolf Witszche also believes that a catastrophic fall in the Sun`s output will have occurred by the year 2050. This German scientist has looked at the data from the Ulysses mission and other sources. Witzsche believes that a cut off point of no solar recovery will be crossed around the year 2040.

After the year 2040, the projected graph of diminishing solar heliospheric wind shows that our Sun becomes comparable to a white dwarf star in output! Of course he does not really mean that our sun will become like a white dwarf star as this scenario is not expected for another six billion years when our Sun runs out of fuel! However this scientist thinks that the sun`s strength will diminish and become very weak.

This may sound very fanciful and fantastic even, and yet there is plenty of evidence that our Sun may be a variable star from the sunspot cycle. As such there is no reason why its output should not vary and become very weak. It is said that in the very distant epochs of our planet`s history the sun used to shine dimly in the earliest epochs of the Hadean and the Archean eons.

4.5 billion years ago, our Sun was 30% fainter than it is today and Earth should have been frozen solid, but it wasn't. This problem was coined as the "Faint Sun Paradox" by the charismatic astronomer Carl Sagan. It seems that Earth is remarkably resilient and that life will survive the harshest conditions (Sagan and Mullen 1972).

A fainter sun may seem an outlandish theory but it is consistent with a burgeoning theory called the electric universe. The electric universe theory turns accepted dogma completely on its head. In this type of universe it seems that unusual things happen. These unusual events include our reliable life giving sun switching off its power output so that it becomes so insipid and pale.

Some solar physicists believe that the central nuclear furnace may not burn constantly but fluctuate, in a similar way to a thermostat. The electric universe theory that says our Sun is a plasma star and burns only on the surface. These theorists believe that the Sun is cooler on the inside than on the outside. The electric theory presumes there is no thermonuclear reaction taking place inside our Sun. This might be why sunspots appear dark in the space telescopes, as the core of the Sun is empty.

This is contrast to the conventional theory of how our Sun works, burning hydrogen for another 5 to 6 billion years before running out of fuel. The Sun will then evolve firstly into a red giant and then a white dwarf. Phew! That is a lot more reassuring to hear. 5 billion years is certainly a lot more time for us to enjoy our beloved sunshine than a mere 35 years. A white dwarf analogy is perhaps taking things a bit too far in the attempt to explain the Sun`s predicted diminishing output.

It is deduced that sunspots are dark since there is nothing inside the Sun at all. The Sun might be an empty hollow globe with all the activity taking place only on its surface. This fluctuation may be in response to conditions in the surrounding galaxy. The idea that there is no nuclear reaction at all occurring at the centre of the sun is certainly radical and alarming. This electric universe theory is in stark contrast to the conventional and accepted model of how our Sun works.

This innovative theory of an electric universe could theoretically be used to explain the sudden onset of a major ice age event. The Sun`s output diminishes to such an extent that the Earth almost instantly becomes frozen. This ice age will then last for around 100, 0000 years before the Sun somehow kicks starts itself back into life and the solar output once more increases.

It is certainly the most incredible theory, and if this electric universe theory is correct then mankind could one day see our Sun change in luminosity before our very eyes. One day we may wake up and look at the sky and see that our sun has changed forever in our lifetime. No longer will it burn brightly and with cheery warmth.

Instead there will be a pale ghostly globe that is almost moon like in our eerily cold skies!

According to the electric universe theory, the Sun`s primer fields are simultaneously powering down as unusual intergalactic changes are taking place. Once a threshold is crossed of a diminishing solar output the Sun may not recover its vitality. The point of no return for a recovery of the sun`s magnetic dynamo is projected to be around the years 2035 to 2040. Then we will be tipped once more into our full ice age lasting 100, 000 years.

There can be no doubt that our universe is brimming with electrical phenomena such as sprites. Scientists at the University of Toronto have discovered a strong electric current in space that is two billion light years from Earth. Philipp Kronberg said that this space lightning electric current was found near galaxy 3C303, and is equal to about 1 trillion lightning bolts (New Scientist 2011). Findings such as these do tend to corroborate an electric universe. This idea is not a new crackpot theory and has actually been around since the year 1903 when George Woodward Warder wrote a book saying that the universe is a vast electric organism.

There is no doubt that this electric universe theory stimulates an enquiring mind to look at the universe in a different light. Electric universe theory states that the chaos of Brownian motion produces the high temperatures we see in the solar corona. The idea that our Sun is not powered by fusion from its core, but somehow draws energy from its galactic environment is certainly unusual.

There is no doubt that our universe is full of exciting electric phenomena such as aurora, lightning strikes and sprites which are a cold plasma artefact. Computer simulations have demonstrated that the motion of the spiral galaxy can be achieved through nothing other than interactions of electric currents in plasma. This theory allegedly explains why 95 per cent of the matter is supposedly missing in our universe that is the missing *dark matter*.

However the electric universe theory has been discounted by mainstream science as the concept challenges most of mainstream cosmology, and even gets rid of the big bang altogether! It also tries to get rid of the concept of gravity and so is considered somewhat heretical. There are however reasons to be concerned about our pithy knowledge of the Sun`s workings. There is plentiful evidence that our Sun is a variable star. This group consists of several kinds of pulsating stars all found on the instability strip, that swell and shrink very regularly.

The nuclear reactions taking place in the Sun are supposed to produce neutrinos. According to the standard model the Sun should be emitting vast quantities of neutrinos. However these hypothetical particles have proved elusive and they may be an invention to balance the books for the theory of supersymmetry. The tiny fraction of detected neutrinos do not account for the present solar theoretical model. So we must ask ourselves the question, do we really know what is going on inside the Sun?

A scientist named Ernst Opik thought that ice ages came when the Sun`s core became swollen and did not convect the heat as efficiently to the outer layer. However it is all speculation since we really do not know exactly what is happening in the core. The theory that neutrinos are produced by thermonuclear reactions at the core may be completely wrong. This possibly incorrect model of the Sun could account for the lack of neutrinos detected to date.

Or perhaps the missing neutrinos could be a warning sign that the Sun is slowing down and gearing up for an ice age. The very small amount of neutrinos allegedly detected might imply that the Sun is far less hot than we think. As a result some solar physicists are questioning the conventional theories about the Sun. Or there is the remote possibility that the lack of neutrinos could mean that the Sun is sick and not producing enough reactions in its core. This is a very worrying idea as it might imply the Sun is dying!

To account for the missing neutrinos some astrophysicists have invented the WIMP, which is a weakly interacting massive particle. American physicists believe there in only one WIMP for every hundred billion particles inside the Sun. WIMPS are said to collide with protons bringing heat from the inner core to the outer core, before arriving to Earth. WIMPS can therefore possibly explain how Sun`s heat gets to Earth and they may account for the missing neutrinos.

If we really understood the nuclear fusion processes taking place inside the sun then perhaps we would have mastered thermonuclear fusion by now. This is plainly not the case as the long running projects at ITER (International Thermonuclear Experimental Reactor) and JET (Joint European Torus) in Oxfordshire have shown. So far the hugely expensive project at Culham in Oxfordshire that uses super-heated plasma has only produced enough energy to boil one kettle worth of energy!

In other words the expensive project has so far failed to deliver the dream of fusion power. I attended a very interesting meeting once at Culham with my friend an Italian plasma scientist. The project is a good example of international collaboration amongst scientists and no doubt some interesting findings are being made along the way.

An Oxford friend, who is a great nephew of science fiction writer H. G. Wells, gave me his paper to bring to the event since he was unable to attend. It was his original theory of cold nuclear fusion! Yes such things really do happen in the city of the dreaming spires! Against my better judgement I stood up with the paper and spoke in front of a wide audience of nuclear physicists!

I read aloud a brief description of the flaws in the current prototype fusion reactor. Afterwards I was immediately circled by a group of serious looking gentlemen in suits wanting to see the full unpublished paper! I kept it safely hidden in my pocket and later deliberately lost it as it had occurred to me that if there was any merit at all in the fusion theory it could be used to make a weapon or fusion bomb, and I am a CND pacifist!

Seriously though, I do believe that the present day approach to the fusion process must be wrong. This may be due to the fact that it is based upon an incorrect model of how the Sun works. The project at JET is creating superheated plasma that is hotter than the Sun! This is clearly a foolish and not to mention a vastly expensive approach. The energy output achieved so far from the expensive experiment is just about enough to boil a single kettle as the joke goes. This makes for a very expensive cup of tea or coffee!!

Just a smidgeon of common sense will tell us that we need to explore a cheap economic power source and the present nuclear fusion experiment is not able to provide this. If we are heading for an ice age of some description we need to expand our power supply as our population increases and more new homes are built. In case anyone is confused at this point let me reiterate that nuclear *fusion* power is preferable to nuclear *fission* power. Fission power can lead to dangerous radiation leaks as happened in Chernobyl, Windscale, Fukushima and Three Mile Island to name just a few nuclear power accidents. Nuclear fusion power which is considered safer has been the Holy Grail quest for energy for a long time now since the 1950`s.

At one point a lot of excitement was generated when two scientists became convinced that they had accidentally stumbled upon cold fusion. In 1989, chemists Stanley Pons and Martin Fleischmann made headlines with claims that they had produced fusion at room temperature. The chemists even published a paper about the accidental discovery. The chemists claimed that their test tube experiment had produced anomalous heat and that they had found small amounts of nuclear by products such as neutrons and tritium. It must have been quite an exciting day in the lab!

Nonetheless the table top experiment was deemed to be a fluke and many scientists doubted whether any nuclear by products had really been detected. Countless attempts to replicate the exciting finding since by scores of hopeful scientists have all proved fruitless. It looks as if the cold fusion reaction incident was a one off and the idea of cold fusion has since been relegated to the archives of oblivion. So once again the focus turned to hot thermonuclear fusion. This also looks like another red herring and has been similarly unsuccessful.

The researchers at JET claim that higher outputs of energy are being achieved recently and that they are slowly on the road to success. This belief seems somewhat overly optimistic. Jet is planning to use Tritium and deuterium plasmas for 2018. The reactions taking place on the Sun may have intergalactic influences which may be impossible to replicate on this Earth in a humble Torus shaped tokamak.

The world`s largest tokamak magnetic confinement ring is being built in France. This project is known as ITER (International thermonuclear Experimental Reactor). It will be finished in 2020 and is hoped to produce 500 mega watts of energy. This fusion reactor is being constructed in a French town named Saint Paul les Durance in Cadarache. China is also experimenting with this concept and is working on a China Fusion Engineering Test Reactor (CFETR).

Really at this stage we should be looking at projects that deliver gigawatts worth of energy as opposed to mere paltry megawatts. I am tempted to suggest that a cyclist pedalling on a self sufficiency type hybrid bike would produce more useful energy than these astronomically expensive attempts at fusion power. Incidentally a new bicycle pedal powered gadget has been invented for an Indian company by Manoj Bhargava. This bike can provide 24 hours of electricity for a rural household.

This device could have a major impact in Indian rural areas that are not supplied with electricity. These bikes are known as free electric bikes. It is hoped that they will make a vast difference to the lives of many who have no access to electricity. It is planned to distribute thousands of these bikes in India. This concept shows how microgeneration of energy can be achieved in a way that creates no pollution whatsoever (Derek Markham 2015).

The microgeneration approach to energy production would not be a feasible means to provide the large scale amounts of electricity needed to run industry and big cities such as London. I can hardly imagine our City bankers pedalling furiously instead of sitting at their desks while making deals with stocks and hedge funds! Still who knows what the future holds? Perhaps one day this may actually happen if there is a national power shortage as predicted by OFGEM.

Our energy hungry society needs to explore the larger macrogeneration projects such as JET even if they are so far proving futile. To this end the plasma scientists need to start examining their models to align with those of solar researchers who are questioning how the Sun works. However if the fundamental theory of how the Sun produces heat is fundamentally flawed then it may be impossible to achieve success using superheated plasmas. The fusion reaction may not take place internally in the sun at all but on the surface of the Sun. However other scientists point out that the sunspots may just appear dark in contrast to the surrounding brilliant glare of the Sun. Therefore we have some solar physicists that believe the Sun is cooler internally.

The mainstream view is that our Sun is much hotter at the core as this is where the thermonuclear fusion reaction is taking place. The converse theory of a cooler central core is a highly speculative theory but certainly we should keep an open mind. This theory might also help explain the missing neutrinos that are proving so elusive to capture here on Earth. It also explains why the heat from the sun increases further away from the surface around the corona. This cool core hypothesis aligns with the electric plasma theory of the universe.

The notion that solar magnetic activity affects the climate, appeared long ago. It is known, for example, that a change in the total quantity of the electromagnetic radiation by only 1% can result in a noticeable change in the temperature distribution and air flow all over the Earth. Ultraviolet rays also cause a photochemical effect, which leads to the formation of ozone.

The flow of ultraviolet radiation increases sharply during chromospheric flares in the Sun. Ozone, which absorbs the Sun's rays, is heated and it affects the air currents in the lower layers of the atmosphere and, consequently, the weather. Powerful emissions which can reach the Earth's surface arise periodically during high solar activity. They can move in complex trajectories, causing aurorae, geomagnetic storms and disturbances of radio communication.

In addition, the solar activity affects the intensity of fluxes of galactic cosmic rays. Another interesting theory has suggested that the Sun is presently immersed in an intergalactic cloud of hydrogen and helium. This cloud is coming from the direction of the constellation Centaurus. This is the brain child of Paresce and Bowyer of the Space Telescope Science Institute in Baltimore. The scientists believe the galactic cloud will result in a cooling effect here on Earth.

The astrophysicist Piers Corbyn has studied the influence of solar activity on our weather patterns here in the U.K. He is convinced that his research on sunspots can predict the weather with reasonable accuracy. Piers is also open minded to the electric theory of the sun and is cynical about manmade global warming. Piers Corbyn is a London based meteorologist and astrophysicist who believes that the Sun is the primary influence on our weather and climate.

Piers forecasts weather patterns using correlations between sunspots, space weather and the climate. Interestingly Piers Corbyn is also leaning more towards a global cooling forecast for our planet. This certainly makes sense as the solar data from Ulysses shows a decline in solar wind. Piers is an elder brother of the popular Labour Party Leader, Jeremy Corbyn. Piers is not following the global warming band wagon and like his talented brother, he is not afraid to voice his own opinions and stand up for his beliefs!

A Whitehouse space program adviser believes that the Sun has cooling cycles. John Casey has worked for NASA on the space shuttle programme and he believes that our planet Earth is in a precarious goldilocks zone as far as solar radiation input is concerned. The Earth`s climate in the Northern Hemisphere can easily be disrupted as there is a very narrow margin of error keeping us warm. The space scientist is concerned that the coming decades may become colder due to the Sun going cold. This could lead to food shortages and crop failures in the Northern Hemisphere.

John L. Casey is convinced that the Sun is about to enter a cooling phase soon and has written the books "Cold Sun" and "Dark Winter". Worryingly the space scientist suspects a link between low solar activity and geophysical events such as volcanic eruption and earthquakes. Other scientists have also published papers suggesting a link between low solar activity and geophysical upheavals here on Earth. Certainly there is a feedback loop between volcanic eruptions and cooler temperatures. The suspended aerosols from volcanic ash can diminish sunlight for years.

Astrophysicist Dr. H. Abdussamatov is the head of space research at the Russian Academy of Science's Pulkovo Observatory in St. Petersburg. He is head of the International Space Station's Astrometria project. Russian solar physicist Abdussamatov predicts another ice age within the next thirty years. Dr. Abdussamatov states that we can soon expect the start of the next bicentennial cycle of deep cooling. According to this scientist the solar radiation flux reaching the Earth has decreased since the 1990`s.

This is caused by a decrease of the solar radius and radiative area. In other words he thinks the Sun is shrinking! Abdussamatov feels that the Sun is actually reducing its diameter and says that this will cause a cooling similar to the Maunder Minimum. Dr. Abdussamatov points out that Earth has experienced such cooling cycles five times over the last 1,000 years. This theory ties in with the solar cooling cycles that Casey talks about in his book "Cold Sun".

Some scientists are giving us a bit more time to prepare for the ice age. The ninety year old distinguished scientist Dr. Fred Singer gave a talk recently about the new ice age. His graphs showed that the major ice age might start within the next two hundred years. Our great grandchildren may live to see this event and this gives us time to prepare. After our mini ice age ends around the year 2055, Singer thinks the temperatures might possibly climb again, only to fall off dramatically as the ice age commences.

There is no doubt that our major ice age is lurking just around the corner. One day our descendants may even wake up to see that a land bridge has opened up once more between Great Britain and France! The idea of being able to walk across to France may seem appealing. However one dare not imagine the cataclysmic event that would result in the lowering of the Channel waters to form this land bridge.

The sun was formed five billion years ago and has circulated around the Galaxy ever since. The Sun has crossed one of the spiral arms called the Orion arm 50 times. A million years ago the sun entered a dark lane situated on the edge of a spiral arm. When it exited the dusty arm, the ice age came to an end. This implies that the space dust is facilitating the ice ages. Every time the solar system passes through this dense dusty part of the Galaxy the volume of dust between the Sun and our Earth is denser.

Certainly there is ample evidence that other forms of dust cool Earth`s temperatures. We are definitely at the stage of the arrival of a new ice age and perhaps the space dust theory proposed by Richard Muller helps to explain the onset of our now overdue ice age. The idea that cosmic dust might periodically cause the planet to cool was first suggested by Richard Muller of the University of California at Berkeley and Gordon MacDonald of the University of California, San Diego, in a short article published in Nature.

Kenneth Farley and Desmond Patterson of the California Institute of Technology in Pasadena have found that the amount of cosmic dust arriving at the seafloor also varies on a 100, 000-year cycle. This intriguing finding exactly mirrors the 100, 000 year cycle of ice ages and interglacials. Farley discovered that the influx of cosmic dust increased suddenly a million years ago roughly the time when the recent wave of ice ages began. It is possible that a bolide collision in the asteroid belt created a large dust cloud, which somehow triggered the first ice age.

It is scientifically accepted that here on planet Earth volcanic ash cools the skies. Sulphur aerosol pollution from volcanoes leads to global dimming and a drop in temperatures. This topic of the global cooling effect of volcanic eruptions will now be explored.

Chapter 5: Volcanic Armageddon.

If a supervolcano erupted today it would unleash a global climatic Armageddon. In the past such eruptions may have resulted in mass extinctions here on Earth. If a large supervolcano such as Yellowstone were to erupt it would unleash a nuclear volcanic winter effect with the dust blocking out the sun for decades. A new ice would arrive within years.

There is also the possibility that our declining magnetic field will facilitate the eruption of our supervolcanoes. Oceanographers have found evidence in deep sea cores that during past magnetic reversals there was a massive increase in volcanic activity. A paper was published in 1970 by Nature magazine by oceanographers J. P. Kennet and N. D. Watkins.

The oceanographers` research showed a mass extinction occurred in the south Pacific which also coincided with a magnetic reversal and ferocious volcanic activity. So just as the magnetic poles were shifting the volcanoes went beserk! Indeed it is thought that Yellowstone last erupted during a magnetic reversal. Therefore we should be keeping a very close eye on our magnetic field which is declining and shifting rapidly. This could be the silent smoking gun that triggers the overdue eruption at Yellowstone.

A supervolcano is defined as having the capacity to produce a super-eruption that ejects at least 1000 cubic kilometers of volcanic material. As well as the climatic effects there is danger from the molten lava and ash. The residents of Pompeii were killed and mummified by waves of fast-moving hot ash and volcanic gasses when Mount Vesuvius erupted in 79AD.

The last known supervolcanic eruption was believed to have occurred about 74,000 years ago at the site today of Lake Toba in Sumatra, Indonesia. It caused a volcanic winter that blocked out the sun for between six to eight years, and resulted in a period of global cooling lasting a thousand years. Toba is a 1,080-square-mile caldera supervolcano in North Sumatra, Indonesia and it is the only supervolcano in existence that can compare with the size of Yellowstone.

When Toba erupted it ejected several *thousand* times more material than erupted from Mount St. Helens in 1980. Some researchers think that Toba's ancient super eruption and the global cold spell it triggered might explain a mystery in the human genome. Our genes suggest we all come from a lineage of a few thousand people just tens of thousands of years ago. This theory could be true if only a few small groups of humans survived the freezing hostile cold years following the Toba eruption.

There are at least seven supervolcanoes that threaten the future of humanity, including Yellowstone in Yellowstone Wyoming National Park. If Yellowstone erupted 87, 000 people would be killed immediately. On February 19 2018 Yellowstone was hit by ten small earthquakes in one day! All of the mini quakes were by Maple Creek. Seismologists said that the site is under severe strain. A swarm of quakes have been recorded on the Western side of the National park. If Yellowstone blows two thirds of America will be uninhabitable and the ash could precipitate the ice age by blocking out sunlight. Still sure you want to plan a sightseeing trip to this stunning National Park?

Valles Calder is another supervolcano lurking beneath a 175-square-mile Caldera in the middle of northern New Mexico. It last exploded 1.2 million and 1.6 million years ago, blasting ash as far away as Iowa. As with other calderas, there are still signs of heat below. Hot springs are still active around Valles. Geologists think Valles caldera is caused by the United States' portion of the North American tectonic plate is being pulled apart.

One of the most fearsome calderas in the world is the 150-square-mile Aira caldera in southern Japan. This supervolcano caldera is by the city of Kagoshima. 22,000 years ago 14 cubic miles of material burped out of the ground and formed the Aira caldera, which is now largely Kagoshima Bay. That is equal to about 50 Mount St. Helens eruptions. The Sakura-Jima volcano has been active on and off for the past century and still causes earthquakes today, indicating that the caldera itself is far from sleeping.

Other danger volcanoes waiting to blow are Mount Fuji and Mount Unzen in Japan. In Columbia there is the Nevado del Ruiz, and the Mauna Loa in Hawaii. Iceland of course has many active volcanoes. Let us not forget either that more volcanoes are found beneath the oceans than above. The supervolcano Uturuncu in Bolivia has been swelling since 1992 and may be preparing to blow. New Zealand is home to the Taupo supervolcano. This scenic visitor attraction is far from dormant and the bubbling hot springs are showing increased vigour.

The Long Valley caldera in California has also been rsiing in the recent decades. This supervolcano erupted 760, 000 years ago and could also be due for a big one. In 1980 there was a 10-inch rise of about 100 square miles of caldera floor. In the early 1990s, large amounts of carbon dioxide gas from magma below began seeping up through the ground. The gases killed trees in the Mammoth Mountain region of the caldera. When these sorts of signs are present, it could mean an eruption is on the way, say volcanologists.

One of the potentially most dangerous eruptions is that of the Cumbre Vieja in the Canary Islands in Spain. The Cumbre Vieja is an active volcanic ridge on the volcanic island of La Palma. If Cumbre Vieja blows it will collapse into the sea causing massive waves and tsunamis. Such a massive volume of debris would displace the water leading to an instant tidal wave. These tsunamis would race across the seas flooding the west coast of Cornwall and Wales. The computer simulations also predict that tsunamis 80 feet high would surge into the East Coast of America!

The gigantic tidal waves would race across the Atlantic Ocean and drown New York City and other cities all along the East Coast. There would be no warning either and no time to evacuate! The 500 feet high tsunamis would arrive in waves and at terrible speeds! The imminent collapse of the Cumbre Vieja is a ticking time bomb that is recklessly being ignored by the British and the American governments. There are many signs that the Cumbre Vieja is chronically unstable and many scientists believe it is about to fail and disintegrate into the sea (Day et al 1999).

Recently in October 2017 a wave of low magnitude earthquakes were detected by seismologists coming from the west flank of La Cumbre Vieja. This might be the precursory activity that is warning us that the sleeping giant is getting ready to blow! There is no doubt that both England and America are in the firing line for a wave of devastating mega tsunamis if this volcano blows itself apart. It is very worrying that volcanic activity is predicted to increase as our magnetic field continues to weaken.

As well as the loss of life from a gigantic tsunami there is a wide body of literature showing the link between volcanic eruptions and cooler temperatures. The sulphur aerosols from volcanic eruptions hang in the skies for a few years like a veil. These aerosols reflect sunlight back into space causing a dimming effect. Even though volcanic eruptions contain greenhouse gases such carbon dioxide the net effect is one of cooling. This cooling effect lasts until the aerosols eventually fall down to Earth as acid rain.

Sulphur dioxide combines with water to form an acid rain. Sulphur dioxide gets converted into sulphuric acid, which then forms aerosols high up in the atmosphere that also serve to block incoming solar radiation for several years after the eruption. Acid rain is harmful to trees and plant life. There have been numerous eruptions in the twentieth century that have had an impact on incoming solar radiation. These include the volcanic eruption of Mount St Helens in 2008. It produced a dimming plume of ash 14,500 metres high into the atmosphere.

Eyjafjallajökull is a volcano on the Eastern Volcanic Zone in southern Iceland that began to erupt on 14 April 2010.A plume of volcanic ash was ejected several kilometres into the atmosphere potentially causing a hazard for aeroplanes. UK airspace was closed causing significant disruption to travellers. The engines of planes can become dangerously clogged with ash from volcanic eruptions and cease to function. The disruption to air travel continued when increased explosive activity at the volcano coincided with northerly to north-westerly winds that brought the ash towards Europe.

The volcanic eruption of Mount Tambora, Indonesia, in 1815 was the largest ever recorded and the resulting ash cloud in 1816 gave rise to what is known as "the year without a summer". Fortunately, modern aircraft did not exist back then, as the disruption would have been unimaginable. Were such a huge eruption to occur today, our modern aviation system would probably be put out of action for months or longer.

The ash plume contains large amounts of microscopic particles of volcanic rock, which can clog up aeroplane engines. Volcanic ash clouds are made up of small abrasive particles that can clog up jet engines that can stop them working. Planes are either re-routed or grounded when there's a danger of flying into ash clouds to ensure the safety of passengers and avoid very costly damage according to Dr David Kerridge, who is the head of Earth Hazards in Edinburgh.

The year 1816 had diabolical temperatures and snow fell in June. It rained constantly and lack of sunlight throughout the Northern hemisphere. Crops failed and many countries endured famine, poverty and sickness. Mount Tambora unleashed an eruption worse than Krakatoa, Washington's Mount Saint Helens, and even Pompeii's Vesuvius. Tambora registered a VEI-7 on the Volcanic Explosivity Index, a metric that measures the size of volcanic eruptions on a scale from VEI-0 (non-explosive) to VEI-8 (megacolossal).

Krakatoa measured a VEI-6, while Mount St. Helens and Vesuvius both rated a VEI-5. When neighbouring volcano Krakatoa erupted in 1883, a huge bang was heard all the way across the Channel in England. French astronomers from Montpelier observed a twenty per cent decrease in solar radiation as ash hung in the skies for three consecutive years. Krakatoa poured 50 million metric tonnes of ash into the skies creating clouds of vitriol.

Giant waves reached heights of 40 m above sea level, hurling ashore coral blocks weighing as much as 600 tons. At least 36,000 people were killed, by the giant sea waves, and 165 coastal villages were destroyed. When the eruption ended, only a third of Krakatoa, remained above sea level. New islands of steaming pumice and ash lay where the sea had once been.

When Tambora erupted in 1815, North American residents reported heavy snow falling as late as the middle of June in 1816. This was the famous American year without a summer. If we were to experience an eruption like Tambora in modern times, the results would be catastrophic. The global population has dramatically risen by billions of people over the past 200 years. The consequences of a super eruption occurring in modern times would be devastating.

Activities like air travel would grind to a halt as volcanic ash can seize up jet engines and cause planes to crash. The global climate change would result in outbreaks of famine and disease.

Experts warn of 90,000 immediate deaths and a 'nuclear winter' across the US if Yellowstone supervolcano erupts. It could release a thick layer of molten ash 1,000 miles away from the National Park. It would be 1,000 times as powerful as the 1980 Mount St Helens eruption'. A haze would hover over the United States, causing temperatures to drop. A supervolcano in the heart of America's northwest has the potential to blanket the continent in ash and create a nuclear winter. A Yellowstone eruption would be one thousand times as powerful as the 1980 Mount St Helens eruption!

Scientists say that we cannot rule out the possibility that an eruption may soon take place. The fact that a major ice age is now due may even coincide with this volcanic eruption. When Yellowstone in Wyoming last erupted it threw more than 1,000 cubic kilometres of ash and lava into the atmosphere. This is 100 times more than the Mount Pinatubo eruption in the Philippines in 1982, which caused a noticeable period of global cooling.

Professor Bill Mc Guire is an expert on natural hazards. He speculates in his books on natural disasters that this volcano may not be dormant and therefore could erupt at any time in the near future. The British Professor of geophysical and climate hazards at UCL is convinced that Yellowstone is now due to erupt. He even thinks that an eruption is quite likely to happen soon according to past records! Mc Guire does not agree with other vulcanologists that the caldera is dormant either.

Three super-eruptions at Yellowstone appear to have occurred on a 600,000-700,000 year cycle starting 2.1 million years ago. The most recent took place 640,000 years ago and this data suggesting

Yellowstone is overdue for an eruption according to Bill Mc Guire. Worryingly some scientists speculate that low solar activity may also trigger an increase in seismic activity and volcanism and there are recent measured signs of a decrease in solar winds. Could a super eruption be the final catalyst for our descent into our massive ice age lasting 100, 000 years?

There is no doubt that a massive eruption would plunge us into an instant ice age as the ash laden skies become dim and dust particles block out the sunlight. The British climatologist Hubert Lamb was the first scientist to develop a Dust Veil Index to measure the cooling effects of atmospheric dust. Lamb studied the effect of volcanic dust on the Earth`s energy balance. This dust index was first developed in the year 1970 to quantify the exact effect of volcanic eruptions on our climate. Later this seminal Dust Veil Index was replaced by the Volcanic Explosivity Index that measured the power of eruptions and consequent effects on the climate.

Our return to a grand ice age is now due. The Earth`s axial tilt has decreased which means it receives less solar radiation. Not all geologists are very impressed with the orbital theory of ice ages however. Some scientists feel that aerial dust might be the missing factor that is involved with initiating ice ages. Dust in the atmosphere also stunts tree growth. This results in narrower tree rings for the years of high dust that is caused by volcanism. The width of tree rings can tell us a lot about the climate for those years of ring growth.

The science of dendrochronology shows how volcanic dust both cools and inhibits plant life. There has to be another causal factor for the overdue major ice age apart from the minor seasonal insolation variables induced by Milankovitch cycles. There can be no doubt that many volcanoes simultaneously erupting would lead to a global cooling event. An ice age climatic tipping point could well be a major volcanic eruption such as that of Yellowstone. This eruption might even be triggered by a pole shift since magnetic reversals are thought to encourage seismic activity.

The ever bubbling Grand Prismatic hot spring attraction in Yellowstone National Park is among the park's many active hydrothermal features created by the supervolcano. The volcano at Yellowstone National Park in Wyoming and Montana sits atop a huge reserve of molten rock. The walls of the Grand Canyon of Yellowstone are made from lava and rocks from a supereruption some 630,000 years ago. There are signs that the volcanic underground activity is increasing towards the State of Idaho and therefore Idaho could be site of the next major eruption! The resultant ash cloud would dim the sun for years and even affect the United Kingdom.

If Yellowstone blew then most of the Northern Hemisphere would be plunged into a catastrophic ice age and crops would fail. An eruption of such a powerful supervolcano would plunge the planet into an ice age. A major volcanic eruption could be this secret smoking gun though some experts say that there is only a one in 700,000 annual chance of a volcanic eruption at the site. Let us hope that this optimistic prediction is correct!

Recently the physicist Michio Kaku announced that the ground has risen by ten inches near the caldera. He suspects that the volcano may erupt soon since it may erupt on a 630,000 year cycle that is now due. During the last 50 years the caldera has been rising at rates comparable to other active and dangerous volcanoes. There may be a link between volcanic eruption and a magnetic reversal. Major volcanism occurred at the Delta Magnetic reversal event 630, 000 years ago around the same time that Yellowstone erupted (Champion 1988).

Yellowstone Volcano's next super eruption will spew vast quantities of gas such as sulphur dioxide, which forms a sulphur aerosol that absorbs sunlight and reflects it back into space. The resulting climate cooling could last up to a decade. The temporary climate shift could alter rainfall patterns. Severe frosts would cause widespread crop losses leading to famine. But a Yellowstone mega blast would not wipe out life on Earth. There were no mass extinctions after the last three enormous eruptions, nor have other super eruptions triggered total wipeout extinctions in the last few million years.

Recently here have been strange sightings of alien lights and supposed UFO`s hovering above Yellowstone. Earthquake lights are known to appear at times of tectonic stress. Hovering globes of light have been seen before major earthquakes. Eye witnesses have reported several hovering glowing orbs that resemble extraterrestrial space ships! There are now many sightings of these strange UFO like lights hovering!

When nature stresses certain rocks, electric charges are activated, as if a battery is turned on in the Earth's crust. The charges can combine and form a kind of plasma-like state, which can travel at very high velocities and burst out at the surface to make electric discharges in the air. Those discharges make the magical alien UFO light shows. These lights may be a sign that the volcano is getting ready to blow says Brian Clark Howard (National Geographic 2014).

Are we all going to die if Yellowstone erupts? Almost certainly the answer is no said the optimistic Jamie Farrell, a Yellowstone expert and assistant research professor at the University of Utah. There have been quite a few super eruptions in the past couple million years, and we are still around he says! However, scientists agree there is still much to learn about the global effects of super eruptions. The problem with the research is that these massive outbursts are rare. It seems very likely that if a super volcano such as Yellowstone were to erupt, it could help to trigger an instant ice age.

Within days, a fine dusting of ash could fall across Europe, according to a UK Met Office computer forecast commissioned by the BBC. The computer model predicts how ash would spread following a nine-day June eruption of 1000 cubic km of ash and gas from Yellowstone. The model shows that the fallout from a Yellowstone super-eruption could affect three quarters of the United States.

The greatest danger would be within 1,000 km of the blast where 90 per cent of people could be killed. Large numbers of people would die from suffocation across the country as inhaled ash forms a cement-like mixture in human lungs. The worst of these health effects would not be experienced in Europe where the ash covering would only amount to a dusting. However the ash would dim sunlight globally.

Many people think that lava flows are the most dangerous volcanic hazards, but ash is often the biggest killer. The sulphur particles would combine with water to form acid rain that is deadly to crops and plants. Because supervolcanoes are highly explosive, much of the magma doesn't get a chance to become lava. Instead it is blasted into countless airborne tiny scorching particles of jagged rock and ash. These razor sharp particles would cut the lungs to ribbons leading to agonising deaths in Wyoming and Idaho.

The most wide reaching effect of a Yellowstone eruption would be colder weather around the globe. Volcanoes inject sulphur gas into the upper atmosphere, forming sulphuric acid aerosols that rapidly spread around the globe. Sulphuric aerosols are the main cause of climatic cooling after an eruption. Aerosols in the upper atmosphere would also scatter sunlight making the sky look like a cloudy winter morning all day long. The overcast effect would be akin to a nuclear winter. The skies in Europe would appear red in the days after the eruption.

To predict how the climate may be affected, the experts relied on historic data from the Toba supervolcano in Indonesia about 74,000 years ago and computer model forecasts commissioned from the UK Met Office and the Max Planck Institute in Hamburg. Experts believe a Yellowstone eruption would inject 2,000 million tonnes of sulphur above the Earth's surface. Once there it would take three weeks for the resulting sulphuric acid aerosols to cloak the globe with devastating effects. Global annual average temperatures would drop by up to 10 degrees, and the Northern Hemisphere could cool by up to 12 degrees!

Some geophysicists are convinced that there is a correlation between increased volcanic activity and declining magnetic field strength. Volcanic eruptions also correlate well with a full magnetic field reversal. Major volcanic eruptions took place at the Jaramillo magnetic reversal and at the Brunhes - Matuyama magnetic reversal around 780, 000 years ago (Schnepp and Hradeztky 1994).

It is possible that major volcanic super eruptions might occur as our magnetic field is plummeting and this will bring on an instant ice age. Increased seismic activity could even affect the precarious San Andreas Fault affecting the densely populated California.
The Earth`s magnetic field is presently decreasing at an unprecedented rate and the South Atlantic anomaly is still growing. The European Space Agency ESA reported in 2014 that the magnetic field is now declining ten times faster than previously thought and could be preparing to flip. Magnetic North is now slowly drifting towards Siberia.

If the magnetic field collapses it would wreak havoc on our satellites and power grids. If a large solar storm arrived it could knock out our electricity supplies leaving us at the mercy of an ice age. Likewise the radiation from galactic cosmic rays that seed clouds would be dangerously increased without a magnetic shield .This increased cloud cover would also facilitate an ice age. We should take the link between declining magnetic field strength and volcanism seriously.

The medieval cold weather of the Dark Ages in Great Britain could have been exacerbated by massive volcanic eruptions. Scientists suggest that the 500 year long icy cold period in the Dark Ages was triggered by an unusual episode of four massive volcanic eruptions. This caused the actual skies to appear darker than normal due to ash and hence the term Dark Ages was coined. This increased volcanism coincided with a weakening of Atlantic currents that caused the cool period to persist for centuries (Miller et al 2012).

The study, led by the University of Colorado Boulder and the National Center for Atmospheric Research suggest that an unusual, 50 year episode of four massive volcanic eruptions triggered the Little Ice Age between 1275 and 1300 A. D. The Black Death plague of 1349 then wreaked havoc on a population weakened by the preceding years of cold weather. The medieval cold weather caused crop failure and famine. The weakened population was soon ravaged by a terrible plague. People dropped dead like flies in the streets of London and around half of the population died from bubonic plague.

The cold summers following the medieval eruptions caused an expansion of sea ice and a related weakening of Atlantic currents. According to computer simulations by the University of Colorado Boulder, the ocean currents slowed. This slowing of currents would also have caused a decrease in the flow of the warming Gulf Stream.

The study used analysis of ice and sediment data, with powerful computer climate models and provides new evidence in a longstanding scientific debate over the cause of the medieval Little Ice Age. The Little Ice Age in the Dark Ages was caused by decreased summer solar radiation, and several erupting volcanoes that cooled the planet by ejecting sulphate aerosol particles. These particles reflected sunlight back into space.

"This is the first time anyone has clearly identified the specific onset of the cold times marking the start of the Little Ice Age," says lead author Gifford Miller of the University of Colorado Boulder. "We also have provided an understandable climate feedback system that explains how this cold period could be sustained for a long period of time. If the climate system is hit again and again by cold conditions from volcanic eruptions there appears to be a cumulative cooling effect."

"Our simulations showed that the volcanic eruptions had a profound cooling effect," says scientist Bette Otto-Bliesner, a co-author of the study. The eruptions could have triggered a chain reaction, affecting sea ice and ocean currents in a way that lowered temperatures for centuries (Miller 2012).

It is interesting to note that the Atlantic Meridional Overturning Circulation (AMOC) was thought to have been affected during this cold volcanically active spell. It shows how the feedback loops such as increasing volcanism act synergistically with the seas to further bring down temperatures. This circulating ocean conveyor incorporating the Gulf Stream is vital in bringing warmth to our shores. The Gulf Stream and its impact on our climate will now be discussed.

Chapter 6: The Sluggish Gulf Stream

London is a dynamic hub of commerce and a thriving city of global importance. London is also one of the wealthiest cities in the entire world. Finally London is also is closer to the North Pole than Newfoundland, where temperatures seldom get above zero in January and February. It is tempting to speculate that perhaps one of the reasons that Great Britain is so wealthy and successful may in part be due to its temperate climate.

Could our economic position be placed in peril by a plummeting climate? If Britain is plunged into an ice age will its economy experience a similar fall? London is a vast dynamic hub of business and finance. Could this prosperity decline if our capital city becomes covered in snow drifts as happened in recent harsh winters? When large amounts of snow arrive in our country everything grinds to a halt.

The British Isles are bathed in the balmy waters and air streams gifted to our shores by the Gulf Stream. We tend therefore to forget just how far north we actually are. We are on the same latitudes as freezing snow covered towns in Canada. Yet we remain relatively frost free and frequently enjoy some mild winters. This is mostly thanks to warm westerlies and warm currents from the Gulf Stream.

When our winters become harsh and bitterly cold it is the result of Northern polar air masses or cold east winds from Siberia, colloquially known as the "Beast from the East". Our mild climate owes a debt of gratitude to the beneficent Gulf Stream. The entire East Coast of America from Florida to Newfoundland also benefits from these warm waters. The Gulf Stream enables the waters of

Florida to remain balmy and warm during winter months. There can be no doubt that the American economy also benefits from the warmth gifted by the Gulf Stream.

A Horizon documentary was recently made about the weakening of our Gulf Stream. The programme was called "Britain the new Alaska". The premise of the documentary is that fresh non salty ice melt water is leading to a shutdown of our Gulf Stream. So a global warming may lead to a cooling in Great Britain. It seems that the feedback loops are mostly weighted towards an ice age for Great Britain. Even if it warms elsewhere in the globe, Britain is always in the firing line for an ice age! We cannot escape the ice and snow as all the odds are stacked against us.

This current of water originates in the warm Gulf of Mexico. The Gulf Stream is a part of the thermohaline circulation of waters sometimes known as AMOC. The thermo refers to the warmth and the haline refers to the saltiness of the water. The thermohaline circulation is mainly triggered by deep cold water masses in the Atlantic and warm water masses in the Southern Ocean. It is also caused by differences in salinity or the saltiness of the water. These differences propel the waters to travel in a constant loop hence the term Atlantic Meridional Overturning Circulation, (Rahmstorf 2003).

A part of the thermohaline circulation branches off to become the Gulf Stream. The benefits of the Gulf Stream cannot be overestimated. It is said that this warm current brings us the equivalent heat of up to one million power stations. More conservative estimates put it at around 27000 power stations worth of free heat! Either way a substantial amount of free energy and warmth is brought to our fortunate shores and to Northern Europe. No doubt this extra energy has also helped our economy to prosper.

Recently there have been plans to directly harness this free energy source using underwater turbines off the coast of America. Using the temperature difference between the warm and cold water it may be possible to turn it into hydrothermal power. Certainly there is a lot of potential energy that is waiting to be harvested.

Were the Gulf Stream to slow down completely there would be a substantial heat and energy deficit. Homes would need more heating to compensate for the drop in temperatures. A study published in Nature indicated that the Gulf Stream or North Atlantic conveyor has slowed by 30%. If it were to fail it could cause another ice age in Northern Europe. This scientific study showed that the Gulf Stream has slowed by 30% since 1957. If it stopped altogether temperatures would fall dramatically in the countries warmed by it (Bryden et al 2005).

This deficit would have to be made up for by increasing our energy supplies. Instead of preserving our vital energy infrastructure we are embarking on a path of power station closure in Great Britain. This may turn out to be the worst policy ever implemented by any Government! During some of the recent unusually freezing cold winters experienced in the U.K. the coal powered Drax power station in Yorkshire was working flat out to keep the lights on as power use surged. Without this hardworking power station, people would have frozen, and the already high winter mortality rate would have risen. Yet our life saving coal power stations are all marked for closure in a policy designed to stave off global warming!

Meanwhile there are worrying signs that the rate of flow of our Gulf Stream may be decreasing. Bill Turrell is an oceanic scientist from the Marine Laboratory in Aberdeen. The team is based in the Faroe Islands off Scotland. Turrell has found a significant drop in the level of saltiness in the water off the coast of Scotland. His team were very surprised at the steep decline in salinity and the graphs showed a precipitous decline in saltiness of the water.

This drop in salinity could lead to a slowing down of the AMOC. There is evidence that the current has slowed due to fresh meltwater entering the thermohaline circulation. This fresh glacial meltwater is diluting the saltiness. This dilution caused by cold fresh water then weakens the current. This dilution then slows the Gulf Stream until it eventually stops altogether.

Scientists now believe that we are getting dangerously close to this cut off point. The unvoiced question is whether the AMOC (Atlantic Meridional Overturning Circulation) currently known to be in decline, could drop off suddenly. This event was depicted in the film "The Day After Tomorrow". This scenario would cause temperatures to plummet. Computer models and evidence collated by Stefan Rahmstorf of the Potsdam Institute for Climate Impact Research in Germany suggests that global warming could turn off the North Atlantic Drift, causing temperatures in northwest Europe to drop by 5 °C or more.

Bill Turrell has found evidence that fits in with Rahmstorf's predictions. He analysed more than 17 000 measurements of seawater salinity between Shetland and the Faroe Islands since 1893.Turrell found that in each of the past two decades the salinity of the deep water flowing south has dropped by 0.01 grams of salt per kilogram of seawater. So its density has probably also decreased by 0.01 kilograms per cubic metre per decade.

"This is the largest change we have seen in the outflow in the last 100 years," says Turrell. "It is consistent with models showing the stopping of the pump and the conveyor belt."

In the 1950s the salinity of the outflow was so stable it was used to calibrate equipment; but now things are changing. Bill Turrell said that this was the most dramatic data he had seen in his entire career. This data has been confirmed by data sets taken by other researchers. However there is some speculation that this measured drop in speed and salinity may be a periodical and cyclical process and that the circulation may recover again (Jackson, et al 2016).

Danssgaard Oeschger Events are recorded temperature changes in the thermohaline circulation (Dansgaard et al 1993). These are dramatic ocean temperature swings that have been recorded in ice core samples. The reader may recall that the graphs show a rapid decline into the start of the ice age and then later the graph shows rises of temperature in the ice age itself. These swings in the Atlantic Ocean temperature became known as Dansgaard-Oeschger events. These are abbreviated to D-O events for simplicity.

These events take place roughly every 1470 years. It seems that the melting ice caps and freshwater from glaciers disrupts the thermohaline circulation. This freshwater reduces the salinity and so slows down the thermohaline circulation. This leads to cooler temperatures in the North. However the Southern Hemisphere may not be affected as it is a bipolar effect. It seems that when the Northern Hemisphere warms up, the ice melts and the AMOC slows down. This in turn cools the land masses such as Great Britain. Therefore these D-O events will likely cool Great Britain.

There is speculation as to what causes this ice to melt and the Sun is one factor. Even during previous freezing ice ages there is evidence that now and then the Sun woke up so to speak and warmed up. The periodic nature of our Sun was discussed in the previous chapter. The awakening of our Sun could have resulted in ice caps and glaciers melting and more freshwater being released in the oceans. This is known as a Dansgaard-Oeschger event. These D-O warming and cooling events have been occurring during our present interglacial the Holocene as well as during the ice ages (Bond et al 2001).

A paleoclimatologist called Hartmut Heinrich discovered that marine sediments provide evidence of large swings in ocean temperatures. Large rises and falls in temperature have been deduced from marine sediments. These sediments provide evidence that cyclical climatic processes are taking place in the Atlantic Ocean currents. It is likely that these changes in Ocean temperature are induced by solar cycles. There are periodic falls and rises in the North Atlantic temperature throughout the last ice age (Elliot et al 2002).

This phenomenon may also occur during an interglacial. Evidence of eight D-O events has been found from ice core sampling as well as marine sediments (EPICA 2004).

The last ice age was characterized by rapid and hemispherically asynchronous climate oscillations, whose origin remains unresolved. Ice cores from Greenland and Antarctica show that sudden temperature shifts, occurred every 1500 years. They were also out of synch in the two hemispheres. When it got cold in the north, it grew warm in the south, and vice versa. Scientists have implicated the culprit behind these bipolar seesaws as the AMOC.

This ocean circulation in turn was most likely influenced by our Sun which may periodically wake up or become sleepy to put it in metaphorical terms. There can be no doubt that the Atlantic Ocean plays a central role in abrupt climate change (Hand; Henry et al 2016). The AMOC brings warm surface waters north and send cold, deeper waters south. In a matter of decades, temperatures plummeted in the north, as the currents brought less warmth in that direction. Meanwhile, the backlog of warm, southern waters allowed the Southern Hemisphere to heat up.

Heinrich Events:

Heinrich events are a more dramatic and deadlier version of the Dansgaard-Oeschger events. A Heinrich event is a rapid release of huge amounts of icebergs from the Canadian ice sheet. An iceberg armada Heinrich event once discharged as much ice into the Atlantic Ocean as the entire ice cap of Greenland. This Heinrich event raised ancient sea levels by ten meters. The icy water brought incredible amounts of freshwater to the North Atlantic. Because freshwater is less dense than saltwater, it can stall the AMOC, preventing the ocean circulation's engine. This is because the circulation depends on a difference between salt and fresh water to keep moving and turning.

The climatologist James Hansen has written a scary book about Heinrich events called "Storms of my Grandchildren". Hansen writes that man could warm the planet with anthropogenic emissions and cause a catastrophic ice melting Heinrich event. Certainly this a riveting book and thought provoking book. There is no doubt that Heinrich ice melting events could be deadly. As I watched a nuclear missile test in North Korea, the thought crossed my mind that a nuclear blast shock wave might one day cause a major ice melting Heinrich event.

Younger Dryas:

The Younger Dryas ice age event occurred around 12,900 to 11,700 years ago. The Northern Hemisphere fell into a deep chill. The sharp drop in temperature of the Younger Dryas was relatively sudden, and took place in decades. This resulted in a temperature decline of up to 6 degrees Celsius. A drop of only 5 degrees Celsius is enough to precipitate an ice age in the Northern Hemisphere. There was colder weather over much of the temperate Northern Hemisphere during the Younger Dryas.

There is evidence of a major extinction event at the Younger Dryas event 12, 900 years ago as the ice age was coming to an end. The giant ice age mammals such as the giant sloth, sabre tooth tiger and the giant short faced bear died out overnight. Mammoths have been found with food still in their mouths as though they suddenly died. Perhaps the extinction was caused by a comet or perhaps a supervolcano erupted.

This cataclysmic event resulted in the ice age resuming with ferocity even though it was due to end. An eruption or asteroid impact set off the Younger Dryas cool period in the Northern Hemisphere that lasted for 1300 years before the ice age finally ended. It was named the Younger Dryas due to an abundance of alpine flower pollen found (Biello 2009).

The Younger Dryas mini ice age event is thought to have been aided by a decline in the strength of the AMOC which transports warm water from the equator towards the North Pole. The slowing of the ocean circulation exacerbated the colder climate. This decline is thought to have been caused by an influx of fresh cold water from North America into the Atlantic.

The Younger Dryas was a period of climatic change, but the effects were complex and variable. In the Southern Hemisphere, there was a slight warming as this is a bipolar seesaw effect. The Younger Dryas data suggest that we need to keep a close eye on our ocean circulation patterns since the temperature dropped dramatically in a very short time when the current changed. Indeed the transition from the greenhouse to the icehouse was probably like a madhouse!

Professor Wallace Broecker described the Gulf Stream as a big heat machine that could prove to be the Achilles heel of our climate if it slowed. Broecker wrote a book called "Unpleasant surprises in the greenhouse?" in 1987. Without our Gulf Stream we would perhaps lose the equivalent of one million power stations` worth of heat. Even if the amount of heat lost was less than this estimate it would still make a substantial difference to the climate of Great Britain. However not all scientists are in complete agreement as to how much heat exactly the Gulf stream brings to our shores and this is an area of scientific debate.

The data from previous glacials shows the Atlantic Ocean currents exert a powerful effect on our climate. Wallace Broecker has spent his entire life studying the Gulf Stream and he believes that it is vital for our climate to stay warm. The Gulf Stream also brings a great deal of warmth to the East coast of America. There is no doubt that ice sheets are melting since the waters of the Gulf Stream is becoming less saline. The AMOC transports a large amount of heat into the North Atlantic where it is given up to the atmosphere and helps regulate the climate in Europe and North America.

This could be a warning sign of an imminent ice age. If Greenland continues to lose ice it can make the AMOC less saline and this in turn can cool down Northern Hemisphere land masses. It is shown that in the past huge injections of freshwater can falter the finely tuned overturning circulation of the ocean currents. The fresh water then disrupts the salinity of the water which makes the circulation less efficient. It seems to a case of heads you win heads you lose as the odds always seem to favour an ice age.

The major effect of a slowing AMOC is expected to be cooler winters and summers around the North Atlantic. A study, published in the current issue of the *Quarterly Journal of the Royal Meteorological Society*, suggests that the Gulf Stream accounts for around 10 per cent of the winter temperature differences between Britain and Newfoundland, Canada. The scientists found that the reason for Britain's mild weather was twofold.

First, there is a genuine maritime effect of being surrounded by a relatively warm body of water. Second, this maritime influence is bolstered by south-westerly winds bringing a warm air mass from the south. These winds would not blow if the Rockies did not exist, the researchers found (Seager 2002). So perhaps Great Britain`s milder climate is being facilitated by the Rocky Mountains. However we should not rest on our laurels and hope that the Rockies will prevent the arrival of our overdue ice age.

The Atlantic Meridional Overturning Circulation and Gulf Stream will also be affected by low solar activity. If the Atlantic conveyor belt were to shut down even temporarily the land around the North Atlantic would cool down by a massive drop of six degrees Celsius. This shut down can be caused by increasing fresh meltwater since a saltier current is more vigorous than a less saline circulation. Paradoxically this is one way that a global warming event such as the melting of Greenland`s ice sheets could result in a cooling for the Northern hemisphere.

If Greenland ice melts, Great Britain will end up with a cooler climate as a result of this freshwater discharge. Freshwater flux from Greenland is composed of melt runoff from ice discharge and calving icebergs. The amount of freshwater flux from Greenland was relatively stable from the late 1970's to the mid 1990's, and then began to increase. Increased freshwater flux could weaken the Atlantic Meridional Overturning Circulation (Yang et al 2016).

The Cryosat-2 Satellite measures the thickness of Arctic sea ice. However it has been found that the measurements can be out by as much as 25 per cent due to snow salinity. Everyone is concerned about the thinning and loss of Arctic ice. At the moment there is an advance/ retreat pattern of ice formation that is also linked to the seasons. Despite compelling images that Arctic ice is shrinking the most recent data showed that Arctic sea ice increased by 2075 km3 in February 2018. However there is no doubt that Arctic amplification will periodically melt the ice.

So it seems that a warming of the Arctic will ultimately bring cold weather to the British Isles. Every roll of the dice is constantly weighted towards a cold climate for Great Britain! So we can forget all about the grape vineyards that global warming enthusiasts have promised us here in Old Blighty! The obvious fact is glaring us all in the face. We are living in an ice house not a green house as there is a plentiful accumulation of snow and ice on our planet.

If global warming did take off we would be in a genuine ice free greenhouse once more. There would not be a trace of snow nor ice anywhere in the entire globe if the Earth was really warming up. Therefore if it snows it is not because of global warming, but because the planet is still in a cooling phase known as an icehouse! When the recent snowstorms arrived in North America some people blamed the heavy snows on precipitation caused by a global warming of the five Great Lakes. The reason it snows is because it is cold and if the planet was truly warm we would be living in a greenhouse with tropical forests at the Poles. We are living in an icehouse and we can expect a lot more heavy snow in our icy future!

There could be other geological reasons than a presumed anthropogenic warming that may account for the melting Arctic ice.

Could one of these causal factors be submarine volcanism? There are over one million volcanoes beneath the sea and hydrothermal vents are found near these underwater volcanoes. The famous Ring of Fire is 25,000 miles long and consists of more than 450 underwater volcanoes in the Pacific Ocean. Volcanoes at mid ocean ridges account for 75% of the magma on Earth (Speight and Henderson 2010).

Scientists have still not mapped all of the planet`s submarine volcanoes and this especially applies to the Arctic region. In 2011 an undersea volcano erupted off the coast of Oregon that scientists had not known existed. Even the Antarctic is part of the active Ring of Fire with a very active volcano called Mount Erebus. In fact Mount Erebus is *the* most active volcano in Antarctica. The juxtaposition of a warm ocean caused by subterranean volcanism and cold growing ice sheets produced a monster storms at the start of the last ice age.

The geologist George Kukla thinks that the start of our last ice age, the End-Eemian, was a time of terrible storms over the oceans (New York Times 16 Feb.1999). This theory of warm oceans and cold skies producing monster storms is aptly depicted by the blockbuster film "The Day After Tomorrow". It is theoretically possible that such super snow storms arose in previous ice ages. The transition from the warm Eemian interglacial to the last ice age may have been a time of climatic chaos. If the Arctic waters do warm it will lead to evaporation of moisture into the cold air resulting in massive snow storms.

The North Atlantic Oscillation is a system of air pressure that affects the weather. During the months of November to April, the NAO is responsible for much of the variability of weather in the North Atlantic region, affecting wind speed and wind direction changes, changes in temperature and moisture distribution and the intensity, number and track of storms. A negative North Atlantic Oscillation strongly correlates with cold U.S. winters.

The NAO was in a very negative phase during the unusually cold snowy winter of 2009/2010. It had not been in such a position since 1865 when records of it began. This rare phenomenon results in a warmer Arctic but colder continents pattern of weather. So one can see that there is a bipolar effect whereby when the Arctic warms, the cold air is often displaced elsewhere (Overland & Wang 2010).

If the Gulf Stream stopped altogether it would lead to an ice age in Great Britain and Northern Europe in less than three years. There is no doubt that the economy would suffer as our infrastructure grinds to a halt. Think of the chaos in London a few years when unexpected heavy snow disrupted travel and there were desperate cries of "where are the snowploughs?" which almost resulted in a Minister for snowy weather being created!

Parts of North America might also resemble the snowy scenes from the blockbuster film "The Day After Tomorrow" if the Gulf Stream stalls. This film sensationalises the idea of an abrupt climate change. The film shows the sudden shutting down of the thermohaline circulation which forms part of our Gulf Stream. This part of the film is entirely feasible and studies show that this oceanic circulation event has happened before thousands of years ago. However in the film this shutdown of the thermohaline circulation is caused by manmade global warming and is not a natural cyclical event.

The film depicts massive vortexes of super powerful hurricanes known as hypercanes. The Earth may have experience hypercanes in the distant past. Extreme conditions are needed to cause a hypercane such as a large volcanic eruption or asteroid strike. A huge temperature difference between the ocean and the skies is also needed to form a hypercane. A hypercane would have a wind speed of over 500mph or 800km per hour and would extend high up into the stratosphere. Hypercanes caused by an impact bolide have occurred on Earth in the ancient past and may have given rise to extinction events (Kerry et al 1995).

Conventional hurricanes do not extend into this higher level of the atmosphere. The stratosphere is a very cold layer of Earth`s atmosphere. Remember the stratosphere is where the ultra freezing noctilucent polar stratospheric clouds are formed by ice crystals at minus 85degrees Celsius. In the film "The Day After Tomorrow" the massive hypercanes that are hundreds of miles wide, create a vortex that sucks down this freezing cold air from the stratosphere.

This leads to an instant ice age in the film. This film is based on a book by Art Bell and Whitley Strieber about a superstorm that is produced by global warming. However such giant storm cells are unlikely in real life to develop over dry land as they do in the film. Hurricanes usually form over water. However there is no doubt that a hypercane, which is a massive type of hurricane, may have once blown with terrible ferocity over Earth.

An atmospheric scientist from MIT, Dr. Emanuel Kerry, speculates an asteroid impact was able to super heat the oceans. Extreme heating of the oceans would have theoretically been able to give rise to a hypercane. Such a wind force could have flattened the dinosaurs in one fell swoop. The super strong wind would also have stripped the protective ozone layer away leading to a vast influx of deadly radiation hitting the Earth. Scientists also wonder if a major impact may have caused the end of snowball earth or deep freezes in our distant past by throwing up vast amounts of energy that melted the ice.

There are many underwater sources of geothermal heat that could warm the Arctic and melt ice thereby lead to a displacement of cold water into our Gulf Stream circulation. For instance there is the underwater Gakkel Ridge. The Gakkel Ridge is the deepest and most remote portion of the global mid-ocean ridge system. It extends 1,100 miles from north of Greenland to Siberia, beneath the Arctic ice cap. The mid ocean ridge is a volcanically active mountain range that runs beneath the North and South Atlantic Oceans, the Arctic Ocean, the Indian Ocean, and the South Pacific.

Underwater hydrothermal vents are seldom included in climate models. However even with a massive heat input from underwater volcanism it is unlikely that a slowing down of the AMOC would lead to the formation of massive hurricanes funnelling air from the stratosphere. The huge superstorm vortex depicted in this gripping film is also unlikely to form as a result of carbon dioxide induced global warming.

The consequences of the Gulf Stream stalling would still be appalling in ways that we have yet to experience. Professor Wallace Broecker said that the Gulf Stream is the Achilles heel of the ocean conveyor belt circulation. He is very concerned that East Coast of America would freeze if the Gulf Stream stopped. Scotland benefits greatly from the Gulf Stream which allows the west coast to grow warm weather plants. Scotland is also on the same latitude as Hudson Bay where polar bears are found! If the Gulf Stream continues to lose salinity it could bring a new ice age for Europe and Great Britain.

Our oceans act as huge heat reservoirs and the heat is circulated back into the atmosphere via the hydrosphere. The world`s oceans maintain a thermal balance across the globe. According to Fred Hoyle it is possible that the thermal equilibrium could collapse within decades and initiate an ice age. If oceans start to lose heat and energy from the weakening sun, the thermal equilibrium can be affected. Ocean currents may be affected by a cooling inertia causing the Gulf Stream to weaken.

If enough water vapour ices up into shiny diamond like ice crystals then further heat is lost and reflected back into space. Once the ice dust veil is locked into place high in the stratosphere, it becomes self sustaining. The ice veil attracts more and more freezing nuclei in a chain reaction. The water vapour starts to fall more often as snow instead of rain as heat is lost from a diminished hydrosphere. We will also start to see more noctilucent mother of pearl clouds that are made of fine ice particles.

Chapter 7: Noctilucent Clouds, shimmering spectres of doom.

There are many varieties of clouds such as cirrus, stratus and nimbus. This chapter will discuss a rare shimmering cloud formation called noctilucent clouds. Noctilucent means night shining and these clouds appear to shimmer in the twilight sky. In 2016 the media excitedly reported that astonishingly beautiful and rare cloud formations were being sighted over the north of England and Scotland.

The press reported that strange and beautiful clouds were seen hovering over the cold skies of England. These clouds were a rare phenomenon known as noctilucent clouds. These clouds are stunningly beautiful. They are also known as nacreous clouds since they shimmer with the opalescent colours that are similar to Mother of pearl. Noctilucent clouds are found high up in the stratosphere in Polar regions and so are also known as Polar Stratospheric Clouds or PSC`s.

These clouds appear to be a recent phenomenon. There was no recorded sighting of these clouds before the year 1885. Satellite data shows that these high altitude frozen clouds have been increasing exponentially during the last few decades. The Aeronomy of Ice in the Mesosphere, AIM satellite, is dedicated to the study of noctilucent clouds. Intriguingly there has been increased sighting of these clouds in the Northern Hemisphere.

Previously the nacreous cloud formations had only been sighted in Antarctica. The nacreous clouds are also becoming much brighter.

The shimmering iridescent clouds are formed at temperatures of minus 85 degrees Celsius. This temperature is colder than average stratospheric temperatures. The implication of this finding is that the upper atmosphere is becoming much colder and it is also shrinking.

These high altitude clouds are comprised of ice particles. Clouds generally don't form in the stratosphere. Typically, it's far too dry there for ice crystals or supercooled water droplets to develop. Therefore the recent appearance of these stratospheric clouds that have been over the North of England and Scotland is something of a rarity. Usually, noctilucent clouds are only seen over the freezing Antarctic.

These clouds are composed of ice crystals and therefore could be a warning that Earth is getting colder. The late Sir Fred Hoyle thought that ice crystals in the upper atmosphere would facilitate an ice age. During winter, when there is little sunlight in the Polar region, the stratospheric polar vortex strengthens and locks out warmer air. This creates extraordinarily cold temperatures in the stratosphere. Water vapour in the stratosphere then changes to ice crystals.

Nacreous clouds form in the stratosphere, at 70,000 feet. This height is more than twice as high as commercial airliners fly, according to NASA. This is also where the ozone layer resides. It is very dry in the mesosphere layer that is 50 miles up. This makes it all the more difficult for these polar stratospheric clouds to form. Scientists believe that these clouds are involved in reactions that destroy the ozone layer.

Nacreous or polar stratospheric clouds, the rare phenomena adorned the British sky in the wake of Storm Henry in February 2016. Henry battered the northern U.K. with winds up to 60 miles per hour (97 kilometers per hour), causing flooding and caused major power outages.. The cold air the storm blew in also led to the formation of beautiful clouds. Andrew Klekociuk, an atmospheric scientist said to the National Geographic "These clouds reveal extreme conditions in the atmosphere and promote chemical changes that lead to destruction of vital stratospheric ozone."

The type of polar stratospheric clouds (PSC`s) that have been appearing over North England are the type 2 PSC`s. These are even colder than type I PSC`s and only ever form in the extreme cold of the Antarctic. It is very unusual for clouds that form in temperatures of minus 85 degrees Celsius to be observed over the United Kingdom. These super freezing clouds should not be able to form over our clement Island. Noctilucent clouds of this colder type are observed over Antarctica and rarely appear in the Northern hemisphere. Something strange is going on over the skies of Great Britain!

This could be the first important sign that a mini ice age is on its way to Great Britain. The areas of Britain that may be most affected by this coming ice age may be North of Yorkshire which is roughly where the last ice sheets of the major ice age reached. These clouds destroy ozone and this leads to more cooling, since loss of ozone allows more heat to escape into space. Manmade chemicals such as CFC`s destroy ozone and so it seems that we are contributing to a global cooling with our pollution.

Other “greenhouse” gases such as methane and carbon dioxide also cause stratospheric cooling and encourage the formation of polar stratospheric clouds according to the meteorologist Jeff Masters. So the effect of our much maligned greenhouse gases appears to be one of stratospheric cooling rather than warming!

Perhaps people are right to be concerned about these trace gases as if we are heading for an ice age they can exacerbate the problem. This cooling and thinning of our atmosphere has affected orbiting satellites and caused an increased drag towards Earth. Many atmospheric scientists tracking satellites are noticing this thinning of the thermosphere such as Robert Kerr, Director of the National Science Foundation Division of Atmospheric Sciences.

Professor Gary E. Thomas of the University of Colorado Boulder is a world expert on these rare shimmering clouds. He agrees that methane and carbon dioxide could become frozen at the mesospheric layer and that these trace gases will enable these clouds to form. He predicts that such clouds will be increasingly seen often over the skies of America.

Professor Thomas is concerned of the effect that industrial emissions may have on our upper atmosphere. It seems that the dice is weighted towards a major cooling event. The so called green house gases can actually turn into ice crystals and thus act as a cooling feedback. I am starting to think that carbon dioxide is more dangerous in its other less known role as a cooling agent. It is rather ironic that the much feared greenhouse gases may be dangerous icehouse gases instead!

Orbits of satellites are drifting because the ionosphere is becoming 17 degrees Celsius cooler per decade. Our upper atmosphere is also shrinking and is now 3% less dense. Could these be warning signs of our overdue ice age? This theory of mine is given support by the findings of Professor Tom Woods. Professor Woods found a link between a shrinking upper atmosphere and low solar radiation (Woods 2010).

So as the sunspots decrease and our solar wind declines the protective atmosphere shrinks bringing freezing cold outer space ever nearer. As the atmosphere becomes less dense the cosmic rays bombard us, and freezing nacreous cloud formations become more prevalent. We should take this appearance of these rare clouds very seriously. It is a warning that our upper atmosphere is freezing out and the perils of deep space are drawing ever closer.

With so much focus on the global warming hypothesis we may be in danger of ignoring these warning signs of an imminent cooling. There is no knowing what the level of severity of the coming ice age will be. There is evidence that our present interglacial the Holocene is somewhat cooler than previous interglacials in our climatological paleohistory. Therefore it is not unreasonable to theorise that the coming ice age may be more severe than previous ice age glacials.

If this is correct then the cold ice sheets may well extend over the whole of Great Britain and Wales. During the last ice age the massive ice sheet stopped just shy of North Oxford. This serendipitous fact will doubtless please the global warming advocate academics in my home city! Perhaps Oxford is the city of global warming after all!

We should however take the warning of a major cooling very seriously. The atmosphere is giving us a stunning visual display that all is not well in the stratospheric air masses over our tiny Island. Since we are now overdue a major ice age glaciation event it would seem reasonable to infer that the new ice age is slowly dawning. Our atmosphere is shrinking because it is cooling rather than warming and solar radiation is declining.

In January 2017 colder type 2 Polar Stratospheric noctilucent clouds were filmed over Finland. It seems that this once rare phenomenon is becoming increasingly more visible. If we ignore this warning sign that our climate is changing to ice age conditions we will be taken by a horrible surprise one day. There is some evidence that an ice age may not come on gradually over thousands of years. Some studies indicate that an ice age can arrive in a matter of years, weeks or even days.

Ice crystals high up in the atmosphere may also lead to major cooling according to Fred Hoyle`s theory of cold ice dust particles leading to ice age conditions. The late Sir Fred Hoyle theorised that as long as the water vapour in our atmosphere remained warm enough an ice age could be avoided. However if the water turned to tiny ice crystals the albedo effect would be enormous. , The transport of water vapour must cause a precipitation rate of about 50 cm to keep the water vapour temperature above -50 degrees Celsius to prevent ice crystal formation.

For comparison, the present day world-wide average of the precipitation rate is about 80 cm of rain, sufficient to prevent ice crystal formation, but not by a wide margin. If the surface layers of the Atlantic Ocean were to cool to the point where an annual average rainfall of 50 cm cannot be maintained, the consequent formation of an atmospheric haze of ice crystals would plunge the Earth immediately back into an instant ice age (Hoyle 1979).

Therefore we need to be vigilant about any increase in icy stratospheric cloud formations as they could be a warning sign that water vapour is icing up into sparkling crystals. Remember that water vapour is the main greenhouse gas in our atmosphere and not carbon dioxide. Water vapour only functions as a greenhouse gas in vapour form.

If it turns to minute ice crystals it becomes a freezing agent instead. The shining ice crystals in turn reflect the sunlight back into deep space. A terrifying albedo effect starts to unfold that is beyond our control. The diamond dust veil then initiates an ice age. Noctilucent clouds may be stunningly beautiful to look at but they may be the ominous portents of an icy future for Great Britain. We ignore this warning sign at our peril!

Chapter 8: Obscured by Clouds, Deadly Cosmic Rays

Cloud cover helps to cool the Earth via the albedo effect. White clouds reflect sunlight back into space in a similar way that ice reflects the sunlight. The albedo effect helps to cool our planet. Therefore if cloud cover were to increase, the Earth would become cooler. Think of the difference a cloud can make to a nice sunny day. When the clouds arrive the temperature difference can often be felt immediately.

Another way that clouds cool temperature is by locking up water vapour which acts as a greenhouse gas in our atmosphere when it is dissipated. This has a net cooling effect. Joel Norris from the Scripps

Institute of Oceanography conducted research that shows that overall cloud changes since 1952 have had a net cooling effect on Earth (Evan and Norris 2012).

The atmosphere is mostly kept reasonably warm for the survival of humans by water vapour and the Earth`s hydrosphere. At night when the sun goes down the temperatures are kept slightly warmer if there is cloud cover. Hence colder starry nights when there is no cloud cover. However the cloud temperature effect on our climate is complex and overall NASA has found a net cooling effect of increased cloud cover. NASA's Earth Radiation Budget Experiment has identified low clouds as being responsible for 60 per cent of the cooling of the Earth by clouds.

Clouds are very important as so far their effects have not been included in any climate temperature models (NASA facts 1999). Cloud cover is predicted to increase as the cosmic ray flux rises. Why are we discussing clouds in our hypothesis about an ice age Britain? The problem with clouds is twofold. Firstly the IPCC has not included cloud cover in any of its climate models! Secondly cloud cover serves to reduce global temperatures by reducing solar influx and by the light reflecting albedo effect.

There is a predicted drop in solar activity in the next solar cycles. When solar activity drops it has an effect on magnetic field strength. The magnetic field plays a vital role in shielding Earth from the incessant bombardment of intergalactic cosmic rays. Therefore when the magnetic field weakens more cosmic rays will enter our atmosphere instead of being deflected. This increase in cosmic ray input has been hypothesised to enable clouds to form more easily.

Therefore when the Sun starts to go inactive the decrease in the magnetic field strength will allow more cosmic rays to enter our atmosphere. These cosmic rays seed cloud formation. Therefore more cosmic rays equates to more global cloud cover. These cosmic rays are also harmful to health and may contribute to high radiation exposure in air pilots and frequent fliers.

The conventional view of climate scientists, as expressed in the reports of the Intergovernmental Panel on Climate Change, is that most of the warming of the Earth's surface over the last few decades is down to the atmospheric build-up of manmade greenhouse gases such as carbon dioxide.

Henrik Svensmark of the National Space Institute in Denmark believes that the cosmic rays and the sun's fluctuating magnetic fields may also play a major role in both warming and cooling. Cosmic rays are charged particles arriving from space or the sun. The intergalactic cosmic rays have been accelerated to speeds faster than any particles in accelerators here on Earth. Galactic cosmic rays are thought to be caused by distant explosions of supernova.

Svensmark has found a strong correlation between these cosmic rays and Earth`s climate. The Danish group has reproduced the Earth's atmosphere in the laboratory showing how clouds might be seeded by incoming cosmic rays. The team believes that fluctuations in the cosmic ray flux caused by changes in solar activity could play a role in climate change.

His graphs show an exact correlation between cooler temperatures and cosmic ray influx. His team has found that cosmic rays that can influence the planet's climate by cooling the atmosphere. It is certainly a paradigm shift to consider that earth`s climate may be partly influenced by factors from outer space!

Cosmic ray flux is also affected by the Sun's output. When the Sun becomes active cosmic rays decrease and the converse applies. When the sun becomes quiescent then cosmic rays increase. The diagram below charts a steady increase in cosmic rays.

Diagram Credit: Spaceweather.com

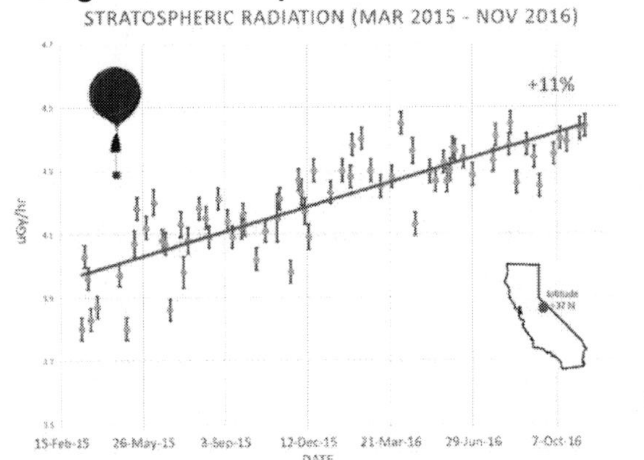

A remarkably good fit has been found between measured increases of cosmic rays and measured decreasing temperatures.

According to the Danish astrophysicist Henrik Svensmark the cosmic rays seed the cooling clouds by ionizing molecules in Earth's atmosphere. The low-lying clouds then have the effect of cooling the Earth further by reflecting incoming sunshine back out to space.

The Sun's magnetic field deflect cosmic rays away from the Earth, when solar activity is high leading to global warming. Conversely, Earth becomes cooler when solar activity is low.

I have recently purchased an excellent book that is selling on Amazon by one of my favourite scientists, the late Nigel Calder. Many will remember Nigel Calder as the editor of New Scientist magazine. Henrik Svensmark and Nigel Calder have written a book about their theory of cooling cosmic rays *"The Chilling Stars: a New Theory of Climate Change"* published in 2007.

Svensmark and Calder have shown by their research that cosmic rays have far more effect on the climate than carbon dioxide emissions. Henrik Svensmark, is Director of Climate Research at the Danish National Space Centre. His studies show that there is a strong correlation between cosmic rays and the formation of aerosols of the type that seed clouds. The authors assert that solar activity is primarily responsible for temperature variation and global climate.

The Sun can affect the amount of cosmic rays reaching our atmosphere via deflection of the magnetic fields. More solar activity leads to fewer cosmic rays. Conversely less solar activity leads to more cosmic rays. The cosmic rays influx then serves to enhance cloud formation. As the sunspot activity lessens, the solar wind diminishes leading to more cosmic rays reaching our atmosphere.

The cosmic rays help to seed the clouds which in turn reflect more heat away from the Earth`s surface. Even if a cloud looks dark grey from below the cloud will always appear bright white from above. The Sun's heliosphere repels intergalactic cosmic rays. The Sun's magnetic field repels the intergalactic cosmic rays from Earth. When the magnetic field weakens, more cosmic rays can enter Earth's atmosphere. There are signs that our magnetic field may be weakening and this could allow more chilly cosmic rays to arrive.

Cosmic rays are ultra- fast proton particles that would threaten life, if not for the shielding of our planet's atmosphere and magnetic field. Cosmic rays create cloud condensation nuclei. Therefore they facilitate cloud formation. Man made pollutants also facilitate this seeding of clouds and so the effect is being amplified. Aerosols in the atmosphere from ships seed clouds adding to the cooling effects from cosmic rays.

Other scientists concur with the cosmic rays affect climate. Professor Jaworowski also believes that the climate is going to get colder rather than warmer in the coming future. He has discovered a close correlation between global temperatures and the intensity of cosmic ray flux. Jaworoski states that solar cycles and cosmic rays influence climate. He is not at all convinced about the role of carbon dioxide (Jaworowski 2004).

There is no doubt that clouds that are induced by cosmic rays have a huge effect on global temperatures. Clouds that appear white on top will provide a reflective albedo effect. It is also interesting to note that the cooling effects of cloud cover have not been included in any of the IPCC climate models! An astronomical study claims to have found a perfect fit between phanerozoic eon temperature and cosmic ray influx. The team studied iron meteorites. They found fluctuations in cosmic ray flux reaching the Earth can explain 66% of the temperature variance over the past 520 million years! (Shaviv and Veiser 2003).

NASA reported that a powerful cosmic ray storm reached Earth in 2016 and there was speculation that it could interfere with mobile phone reception. Cosmic ray storms could be a portent of a new ice age. Reports in spaceweather.com found that cosmic rays are intensifying. Neutron monitors around the Arctic Circle are recording an increase in cosmic rays. Helium balloons launched over California have found a similar increase. Could this increase in cosmic rays help explain recent stormy weather events?

Measurements show that someone flying back and forth across the USA, just once, can absorb as much ionizing cosmic radiation as given by three dental X-rays. Likewise, cosmic rays can affect mountain climbers, and astronauts on board the International Space Station. If our cosmic rays do increase as our ice age approaches frequent fliers may get an extra dose of ionising radiation. The path of the Sun through the spiral arms of the Milky Way exposes the Earth to varying intensities of cosmic rays. This passage of our planet through dusty spiral arms in our Milky Way may also explain the onset of ice ages.

One should perhaps expect climatic variations while we roam the galaxy. The density of cosmic ray sources in the galaxy is not uniform. Cosmic rays are concentrated in the galactic spiral arms. Cosmic rays arise arises from supernovae, that are predominantly the end product of massive stars, which form and die primarily in spiral arms. Each time we cross a galactic arm, we should expect a colder climate or a major ice age. This idea ties in with the cosmic space dust theory of ice ages. Some of the global warming witnessed over the past century could be attributable to the increased solar activity that has diminished the cosmic ray flux reaching Earth.

Tracks of natural cosmic rays have been made visible by a detector at CERN in Geneva. CERN is the European Organisation for Nuclear Research and is home to projects such as the Large Hadron Collider particle accelerator. This underground circular accelerator recently discovered the Higgs Boson or God particle, so called as it is thought to give rise to all matter and mass. Some scientists voiced concerns over the very high speeds being used in the experiment and one even said that it might create a dangerous black hole!!

However CERN would like to see if mini black holes arise, in order to explore new theories about our universe. CERN`s website says that If micro black holes do appear in the collisions created by the LHC, they would disintegrate rapidly, in around 10^{-27} seconds. They would decay into supersymmetric particles, creating events containing an exceptional number of tracks in our detectors, which we would easily spot. Finding new particles would open the door to yet unknown possibilities about how our inverse works. If they created a big bang would a new universe be born which then proceeded to replace our present universe?

CERN has vigorously denied that the strong magnetic fields generated by the accelerator can affect weather or cause earthquakes here on Earth! Neither will a black hole be generated that will proceed to gobble up our planet! The Earth`s magnetic field may play a vital part in the onset of ice ages. This is because a weaker magnetic field can allow more cloud forming cosmic rays to penetrate. CERN may provide useful data to track the incoming cosmic rays.

Feeble solar activity and a weak heliosphere both coincide with increases of cosmic rays. Radio carbon 14 in samples on Earth is used to measure these cosmic rays. There is a theory that the galactic centre of our galaxy the Milky Way might be a source of super powerful cosmic rays. The galactic centre is 23, 000 light years away in the constellation of Sagittarius.

The Voyager Missions measured cosmic rays. Voyager found that these highly energetic rays may be increasing in our solar system. This could be due to our solar system passing through an energetic part of the galaxy. Our solar system is not stationary and moves along and sometimes enters regions of dense interstellar matter. It is possible that a cosmic ray bombardment might originate from the galactic centre in line with the Mayan predictions of a harvesting of souls. It is also thought that a giant black hole lurks in the galactic centre.

The ozone layer may also be damaged by energetic cosmic rays. The Montreal Protocol was enacted to protect the ozone layer and banned ozone destroying chemicals. Other scientists are concerned that rocket launches may be destroying our ozone layer. Professor Darin Toohey of University of Colorado, Boulder, believes that rocket emissions are worse than CFC`s. These chlorofluorocarbons were banned by the Montreal Protocol to protect the ozone layer. However Professor Toohey feels that a few NASA rocket launches will undo all of the good work from banning CFC`s.

A Canadian astrophysicist from the University of Waterloo believes that cosmic rays have caused most of the damage to the ozone layer. He has found a correlation between cosmic ray intensity and the destruction of ozone. Canadian scientist Qing- Bin Lu is convinced that 100 per cent of Antarctic ozone loss is due to the energetic cosmic rays that interact with the chlorofluorocarbons. Therefore the primary instigator of the ozone depletion might be the cosmic rays originating from the galactic centre. I am sure the ancient Mayans would be pleased with this theory that aligns with their theory of the galactic heart!

If an ice age is due one might expect the cosmic rays to increase and the ozone layer to diminish. The ozone layer is vital for life on Earth as without it most plants and crops would perish resulting in mass starvation. A NASA experiment with a basil pot plant found the plant turned brown and died when exposed to ultra violet light without protective ozone.

Without the ozone layer protecting us we would soon become blind from radiation induced cataracts. Staggering blind people would roam the Earth searching for food in a zombie style apocalypse. As the vegetation continued to die the oxygen levels would reduce by half. Without oxygen human life as we know it ceases to exist though primitive anaerobic life forms might survive.

An aerospace engineer claims to have discovered that the ozone depletion is far worse than we are being told. According to the engineer we have only a few years before the ozone layer vanishes! This concern ties in with those of other scientists who fear that our ozone layer is being shredded by rocket launches. It does seem strange that we are now seeing noctilucent clouds in the skies of Great Britain. In 2018 Imperial College London found that ozone is decreasing in lower latitudes in locations away from the Poles. This finding is very worrying. This might explain the increase in the noctilucent clouds that have been observed at lower latitudes.

HESS Observatory, is an international collaborative venture. This group confirms that the galactic centre of our Milky Way is a source of energetic cosmic rays. Think of Hawking radiation escaping form a black hole! Galactic core outbursts are possibly the most energetic phenomenon in the universe. The black hole produces super energetic cosmic rays that are accelerated at speeds 100 times that of the particle accelerator, the Large Hadron Collider at CERN. Cosmic rays have energies of up to 100 teraelectronvolts (TeV). To reiterate, cosmic rays are thought to seed clouds and therefore facilitate snow storms.

The rays are also produced by remnants of super nova and pulsars. If a super cosmic ray bombardment arrived it could help trigger the ice age by seeding massive snow storms. The term galactic superwave was coined by the astronomer Paul LaViolette who is president of the Starburst Foundation. He discovered that galactic superwaves are emitted every 13,000 and 26,000 years. Some scientists believe that in the past huge galactic superwaves have arrived from this region causing massive climate changes on Earth.

Ionisation from intense cosmic rays leads to a depletion of the ozone layer. High levels of ionising radiation in the stratosphere lead to decreased levels of protective ozone. This depletion results in a global cooling. A high volume of cosmic rays seeds clouds and could trigger a snowblitz on Earth. Some scientists speculate that a galactic superwave is already on its way to Earth and may cause a deadly ice age. The idea that a black hole from the galactic centre could trigger an ice age on Earth is certainly very strange and spooky!

Chapter 9: Our Dimming Skies

As our skies become more polluted they facilitate global dimming. The term global dimming was first used by Gerry Stanhill an English scientist who worked in Israel. Stanhill observed a huge drop in solar radiation from the 1950`s to the present time. He published a paper in 2001 showing that he had observed a 22 per cent fall in solar radiation in Israel.

He then replicated this finding by taking measurements from all over the globe. Everywhere he found the exact same result. Stanhill used pans filled with water to observe the evaporation times under daylight. The rate of water evaporation has been slowing down, which is consistent with global dimming and global cooling. The published paper of the new global dimming concept aroused concern over man`s polluting activities (Stanhill et al, Climatic Change 2004).

Man is polluting the planet and we have yet to see whether this may hasten the onset of our major ice age. Once again, as in the days of London pea soupers, our skies are becoming murkier and smog laden. The dire pollution has recently led the London Mayor Sadiq Khan to call for a ban on woodburning stoves and diesel cars in the capital city.

Certainly there are thousands of deaths each year now linked to air pollution. The Clean Air Act of 1956 banned open fires after the deadly smog of 1952 that killed many London citizens. These days the seminal Clean Air Acts are all but redundant as homeowners are encouraged to install these woodburning stoves! As a result thousands are dying from asthma, bronchitis and heart disease linked to the polluted air.

Recently Stanhill`s global dimming water evaporation results have been replicated by Australian scientists who used a different way to measure the incoming radiation. A satellite array around the Earth is recording worrying falls in levels of solar radiation reaching the surface. A shocking fall of 37 per cent was recorded over Hong Kong.

If this dramatic trend of decreasing solar radiation or insolation continues this could facilitate a return to an ice age. The main cause of this accurately measured global dimming is thought to be air pollution, from particulate pollution dirty emissions such as untreated vehicle exhausts and open wood or coal fires.

In 1974 Time Magazine carried a front page article warning that burning too much coal in open fires could lead to a new ice age. This is due to hazy particulate pollution blocking out the sun`s rays. The idea that hazy pollution can prevent sun`s warming rays from getting through has been known for a long time in scientific circles. Indeed before our present day zeitgeist of anthropogenic global warming. many scientists in the 1970`s voiced concern over the imminent ice age.

However we need to recall that the Solar and Heliospheric Observatory, SOHO, mission also recorded a drop in the actual output of the sun by 0.015 per cent. This fall in irradiance was measured by the Virgo instrument. This depicts a huge fall in solar output and might also account for global dimming. The observed dimming has measurable differences across the globe.

The dimming was observed to be more severe in the Northern Hemisphere and over countries like China and India. The skies over Malaysia are very dim due to abundant use of open wood fires for cooking. There is no doubt that any decrease in sunlight reaching the surface will have a negative impact on crop yields. Less sunlight results in less energy for crops to turn carbon dioxide and water into the nutrients needed for plant growth.

The ongoing CLOUD experiment at CERN is providing new data about human activities and their effect on cloud formation. The scientists in Geneva are studying cloud seeding from emitted aerosols from sulphur dioxide and other man made industrial emissions. There is no doubt that humans are polluting the skies. This anthropogenic airborne pollution may lead to a global dimming effect and this topic will now be discussed.

Global dimming is caused by sulphur aerosols that reflect sunlight back into space. Pollutants become nuclei to seed cloud droplets. It has been observed that this seeding of pollutants leads to a greater rainfall over busy cities. Polluting vehicles make the dimming worse. Diesel emissions and burning wood in open fires cause both pollution and dimming.

Diesel vehicles have caused tens of thousands of premature deaths each year here in the United Kingdom. The PM2.5 particulates from diesel are also linked with preterm births worldwide (Malley et al 2017). Diesel cars were given tax breaks by our Government in the belief it would save carbon emissions and prevent global warming. All of the extra pollution from ill conceived Government policies will simply exacerbate the trend for global cooling. NASA recently announced that it had found the most polluted skies over England, Europe, New York, Hong Kong and China from its satellite data.

Very high levels of nitrogen dioxide have been observed and these emissions are very harmful to health. The nitrogen dioxide emissions are coming mostly from vehicles such as the older polluting diesel cars. Diesel cars were promoted heavily in the U.K. as a low carbon and clean alternative! This is one small example of how confused the policy makers are becoming. Diesel was given a tax break for being low in carbon dioxide. Since carbon dioxide, a non toxic gas, was designated as a pollutant diesel was preferred.

Particulate air pollution in Beijing is providing a cooling canopy of air pollution. China has made the transition from rickshaw to motor car very rapidly. The result is a toxic nightmare of pollution. China does not have the advanced coal power stations that remove particulates using chimney stack scrubbers .China needs to do more to remove the burnt by products. Recently China has made progress in doing this. However the main problem seems to the countless factories that burn coal and let the untreated emissions escape.

Such emissions will appear a murky brownish colour and will be full of sulphur dioxide. Opaque sulphur aerosols act as powerful global cooling agents.The sulphur aersosls now polluting China`s skies burn the lungs of Chinese citizens.These aerososls are similar to those produced by volcvanic eruptions. When sulphur dioxide combines with water it produces an acid rain that kills vegetation.Japan is concerend the acid rain might ruin their crops. Japan has even offered to pay for the desulphurisation equipment needed for China`s coal plants!

China is now making big steps to install desulphurisation equipment in all of its chimney stacks but still lags far behind. China may decommission all of its coal plants soon to remedy the pollution problem. This pollution haze is unlikely contribute to global warming however as has been wrongly asserted. Science journalists broadcasting on Channel 4 News announced incorrectly that the smogs will lead to global warming.

Most likely the dense smogs are acting in a similar way to a geoengineered sunshield. This cooling effect has led many climate modellers to try and estimate by how much this sulphur induced cooling haze from China will offset any hypothesised carbon dioxide induced warming. Global dimming, it is hoped, may counteract some of the hypothesised anthropogenic warming. Of course if our ice age is about to arrive, any cooling from the global dimming will make the situation a lot worse!

The concept of a nuclear winter is a good example of a catastrophic global dimming. If several nuclear bombs were simultaneously detonated life would die out, skies would darken, crops would fail and a new ice age would dawn. Pollution particles help seed clouds. Every water droplet in a cloud needs a particle to seed itself around. Ions become attached to particles and act as cloud condensation nuclei. Negative ions, or electrons, collect dust particles as they rise upwards.

The process of cloud seeding has actually been replicated in the laboratory. These particulate seeded clouds then become more reflective than usual. This in turn results in more solar radiation becoming reflected back into cold space. It is of great concern to learn that some scientists wanted to add *more* of these particles into the atmosphere in order to geoengineer a canopy to further cool the globe.

Some climate scientists have theorized that aircraft contrails or vapour trails are implicated in global dimming, but the constant flow of air traffic previously meant that this could not be tested. The total shutdown of civil air traffic during the three days following the September 11, 2001 attacks afforded a unique opportunity to observe the climate when absent from the effect of contrails. An increase in diurnal temperature variation of over 1 °C (1.8 °F) was observed in some parts of the U.S. This finding proves that aircraft contrails lower daytime temperatures and are causing global dimming (Travis, et al 2002).

Aerosols and other particulates absorb solar energy and reflect sunlight back into space. The pollutants can also become nuclei for cloud droplets. Water droplets in clouds coalesce around the particles. Increased pollution causes more particulates and thereby creates clouds consisting of a greater number of smaller droplets. The smaller droplets make clouds more reflective, so that more incoming sunlight is reflected back into space.

In 1971 Dr. Stephen H. Schneider from the National Center for Atmospheric Research in Boulder, Colorado, claimed that pollution would soon *reduce* the global temperature by 3.5°C. The National Science Board of the U.S. National Science Foundation announced that "The present time of high temperatures should be drawing to an end leading into the next glacial age."

As a postdoctoral fellow at NASA's Goddard Institute for Space Studies, Stephen Schneider decided to examine whether the particulates that cool would counteract the carbon dioxide green house effect. In 1971, Schneider was lead author of a paper in *Science* titled "Atmospheric Carbon Dioxide and Aerosols".

This paper used a model to examine the competing effects of cooling from aerosols and hypothetical warming effects from CO_2. The paper concluded that mankind`s injection of particulate matter in the atmosphere could make the air four times more opaque. The team`s calculations suggest the aerosols would result in a *decrease* in global temperature by as much as 3.5 degrees.

Therefore the team decided to favour nuclear power as a future source of energy to prevent the cooling sulphur emissions that might lead to an ice age! These days we have the *exact opposite* scenario whereby some authors favour nuclear power to prevent global warming!

Today in 2018 we have a few green environmentalists who are proposing the rollout of nuclear power. Today we have the authors Mark Lynas and George Monbiot and James Lovelock agreeing that it is better to embrace nuclear power to help the climate stay cool. Both Oxford based authors, have written excellent books about the perils of a warming world. Both authors agree with the premise that nuclear power will help to prevent a runaway warming that burning will fossil fuels induce.

The enterprising Mark Lynas even went so far as to embark on a fact finding mission to Chernobyl. I was astounded to turn on my television one day to see Mark Lynas standing bravely in the desolate Chernobyl landscape. The fearless climate theorist had plonked himself squarely in the centre of the bleak radioactive exclusion zone. The intrepid explorer stood alongside the abandoned buildings and ghostly fairground ferris wheel of this once thriving city. He looked very nonchalant as he stood in the bleak desolate danger zone wearing no protective equipment to shield him from residual radiation.

The reason for this reckless act of journalism soon became clear. Mark Lynas had gone to Chernobyl in order to promote the idea of nuclear safety! The irony of this mission kept television viewers suitably amused. Not for the first time was I appreciating the unabated merriment caused by the entertaining author`s antics! The nuclear power advocate had endeavoured to show his captive audience that the dangers of radioactive fallout were vastly overplayed.

He hoped to prove just how safe nuclear energy is by giving a broadcast from the danger zone! Since Chernobyl had seen one of the worst nuclear disasters in the history of nuclear power this did not seem such a great way to promote the safety of nuclear energy! Somehow this obvious fact seemed to have wafted over the author`s head, just as the cloud of radioactive isotopes had wafted over the United Kingdom in 1986!

The fallout from this cloud was felt for many years especially in the Welsh hills where sheep grazed. Welsh lamb was tainted by the radiation and farmers were pushed to the brink of bankruptcy. The effects lasted for decades and restrictions on the lamb farmed in North Wales have only recently been lifted a few years ago in 2012. Altogether around 5300 hectares of farmland was affected near the National Park in Snowdonia.

The radioactive particles were absorbed by the grassy vegetation and in turn the grazing animals ate this radioactive grass. The lambs had to be scanned for caesium levels before they could be sold. Those lambs grazing on higher ground absorbed more caesium. The beautiful Snowdonia landscape of North Wales had become contaminated. The radioactive fallout also affected Cumbria and parts of Scotland because of the prevailing winds at the time.

This nuclear disaster was not an isolated mishap in the history of "safe" nuclear power. There have been nuclear releases from Three Mile Island, Windscale and Fukushima to name a few that spring to mind. The Three Mile Island incident was a partial meltdown that took place in 1979 in Dauphin County in the United States. A study found that those living within a ten mile radius received radiation levels equivalent of one chest x-ray. It was considered a minor accident and everyone was very relieved it had not evolved into a China syndrome scenario and felt that they had got off lightly.

The Windscale Nuclear power plant disaster in Cumberland Cumbria, took place in October 1957. This was rated as a severe incident and released clouds of radioactive isotopes of iodine. The cloud spread as far as Europe. Eleven tons of uranium caught fire on

October 11 1957.The iodine isotopes may have caused extra fatalities from thyroid cancer.

Our thyroid glands need iodine to function. Therefore the presence of radioactive iodine in the atmosphere causes an uptake of the radioactive form of iodine as the body confuses it with the beneficial trace element that is needed to produce energy. The Irish Sea was also contaminated by the accident and oysters became radioactive (Preston et al 1968).

The politician Ed Davey came to Oxford to talk about his work at the Department of Energy and Climate Change (DECC).He spoke about the work his department was doing to insulate more of Britain`s chilly homes. He also spoke of the cleanup operations at nuclear power plant Sellafield which was formerly known as the Windscale plant. However when I asked him about the budget allocated to clean up Sellafield he answered truthfully that £billions of taxpayer funds were being spent cleaning up radioactive waste.

It seemed that most of the department`s budget was being spent on Sellafield decontamination. It was a moment of revelation as I realised that nuclear power could not be the solution of our energy needs. Surely this was a galling example of the hidden costs of nuclear power. Paradoxically many so called green environmentalists are calling for nuclear to be favoured as a low carbon source of energy. It may be low in carbon but it is still frighteningly high in radioactive isotopes!

The total costs to clean up Sellafield have been projected to total £53bn. The Nuclear Decommissioning Authority estimates that the site will be clear by the year 2120.Yet it seems that we have not learnt a lesson about expensive nuclear power as this country now proposes to build the most expensive nuclear power plant in the world at Hinkley Point. Nuclear power is both hideously expensive and potentially lethal to the environment.

Japan was shaken to its core on March 11 2011 by a huge earthquake that caused a tsunami to rise up and hurtle towards the supposedly protected shores of Japan. The floodwall levees were breached and the waters gushed over land with a frightening ferocity. The magnitude 9 earthquake was so huge that it was reported that the entire globe wobbled slightly on its axis. NASA reported that the Earth had shifted slightly on its figure axis leading to a shorter day by 1.8 microseconds.

Amazingly the effects of the earthquake were felt all over the globe and even affected an ice shelf in Antarctica. The Sulzberger Ice Shelf in Antarctica was shaken by the travelling ripple and large chunks broke off that were the size of a city such as Manhattan. Satellite images showed large chunks of ice floating off into the sea shortly after the magnitude 9 quake. This massive earthquake even sent shock waves into our atmosphere. The effect created electrically charged ripples 220 miles high up in the atmosphere.

The perils of constructing nuclear facilities in an earthquake zone were demonstrated by damage caused to the Japanese nuclear plant at Fukushima. The legacy of Fukushima lives on as a chilling reminder of the dangers of nuclear power. Greenpeace issued a report saying that the contamination from the damaged Fukushima Daiichi plant will last for centuries as it is distributed via the Pacific Ocean through various plant and marine food chains (Greenpeace press release March 2016).

The reason for this digression on nuclear power safety is to highlight the danger of trying to curb carbon emissions. Green authors such as George Monbiot, Mark Lynas and even James Lovelock feel that we must have supposedly "low carbon" nuclear power on the table in order to avoid a catastrophic global warming event. Even the premise that nuclear power is "low carbon" has been called into doubt by a recent study published in the Ecologist in 2015.

The idea that nuclear power might possibly be designated as low carbon was researched by Professor Keith Barnham. The detailed report found that there is no consensus on the carbon footprint of nuclear power and no agreement on the lifetime carbon emissions generated by nuclear power stations. So called "low carbon" nuclear is in fact high in carbon emissions as well as high plutonium. A lot of the carbon emissions are incurred during the construction phase as concrete is held to be a high carbon substance. Therefore it seems that the pursuit of low carbon nuclear is somewhat misguided.

Despite my grave misgivings I do feel that nuclear power would be invaluable during ice age conditions as a lot of energy would be needed. Mankind will need a huge investment in energy infrastructure and power plants to cope with such a drop in temperatures. Although as an environmentalist I am against nuclear power, it may have to an option on the icy table. The Northern Hemisphere may struggle to meet its energy needs and this topic is discussed later in this book.

Did Stephen Schneider make a mistake when saying it would be preferable to favour nuclear power over fossil fuels? Schneider was undoubtedly correct in his assertion that sulphur aerosols act as cooling agents in global temperatures. Let us not forget either that Schneider favoured nuclear power to help prevent the ice age. Schneider later became concerned about anthropogenic warming along with scientists in the IPCC (Intergovernmental Panel on Climate Change). I think his early idea of an ice age was on the right track and that safely maintained nuclear may prove useful during the ice age.

In the mid-1980s, Atsumu Ohmura a geography researcher at the Swiss Federal Institute of Technology, found that solar radiation striking the Earth's surface had declined by more than ten per cent. Many experiments have been carries out worldwide on the evaporation of water from flat pans exposed to sunlight. These experiments all show that the rate of water evaporation is slowing, which is consistent with a global dimming hypothesis.

It could also tie in with a phase of the Milankovitch cycle of decreased insolation as Earth tilts away from the Sun on its axis. Certainly the pan evaporation experiments do not support the global warming hypothesis. Indeed the converse applies and the pan evidence is supporting a global cooling scenario. The decrease in evaporation is consistent with a decrease in sunlight from increasing cloud coverage and global dimming aerosols (*Science* 15 Nov. 2002).

Natural global dimming aerosols are emitted worldwide from volcanic processes. For example there is constant particulate emission from rumblings of volcanoes such as Mauna Loa, the world`s most active volcano in Hawaii. This constant emission of volcanic aerosols consists of both carbon dioxide as well as sulphur aerosols. Indeed natural processes such as volcanic activity do contribute a large amount of atmospheric greenhouse gases.

Yet intriguingly the net effect of volcanic eruptions appears to be one of cooling via a global dimming effect. It is within the realms of probability that a large amount of the present day increase in atmospheric levels of carbon dioxide are the result volcanic emissions from Mauna Loa rather than the activities of man. However we must not be complacent about the deleterious effect of our polluting activities are having on the atmosphere. The endless launch of test missiles is no doubt contributing to the shredding of our vital ozone layer.

Indeed it has been said that of the present day carbon dioxide in the atmosphere, man is responsible for less than one per cent. The rest of the carbon dioxide is derived from organic processes such as decomposition of plants and ocean dynamics. However one has to realise how powerful is the catalyst effect of carbon dioxide. Carbon dioxide is a vital catalyst which enables all life on earth to flourish. Despite being present in relatively minute quantities a trace gas may have a huge effect.

Our planet would be a barren uninhabited wasteland were it not for carbon dioxide. Despite it comprising only 400 parts per million this important gas enables all photosynthesis and plant respiration to take place on our beautiful planet. This depicts how a small part of the atmosphere can exert a very large effect. Therefore it is not surprising that some scientists are concerned about rising carbon dioxide levels. Without this trace gas there would be no vegetation and by logical deduction there would be no life on Earth whatsoever.

Therefore despite being in minute quantities this gas has an enormous effect. The burning question is can such a small increase of this life giving gas from man`s activities make such a difference that we may stall the onset of our looming ice age? This is complex but it is possible that black soot deposits on white snow could counteract the albedo effect. This is the one possible way that man theoretically might be able to warm the planet.

Indeed this idea to geo-engineer the prevention of an ice age with soot was put forward by the scientist Fred Singer. This distinguished scientist realised that an ice age is coming within 200 years at the very latest. Fred Singer therefore proposed that dirty black soot could be sprinkled on the white pristine snows of Canada to absorb the sun`s warming rays. Needless to say most environmentalists would not be pleased at this idea!

Carbon dioxide comes in many forms including dirty polluting hydrocarbons. Some of this dirty black carbon falls as soot on pristine white snow. These dark patches of soot appear dark from space. Since they are dark they can absorb the sun`s rays. This absorption of solar energy serves to counteract the albedo effect. Dirty sooty hydrocarbons pollute our skies and are actually be cooling rather than warming our planet.

So here we see that sooty carbon may either warm or cool the planet and its climatic effects are very complex. In its pure form carbon dioxide is an innocuous harmless gas essential to life. However carbon dioxide comes in combinations such as dirty hydrocarbons that pollute our skies and damage our lungs. Hence it is easy to understand how carbon emissions are now considered a bad thing for our planet`s ecosystems.

Despite widespread concerns, man is still managing to emit vast clouds of hazy pollution especially in countries such as Malaysia where wood burning for cooking is common. The brown hazy smogs can be clearly seen from satellite photos from space. This Asian brown cloud as it is called, covers much of the Indian Ocean.

It is caused by cooking with open fires using wood. This also creates health problems from the smoke inhaled indoors. Two million die in India alone from this noxious particulate laden pollution. The smoky method of cooking is used because of extreme poverty and of households being unable to afford cleaner methods of cooking.

Yet ironically woodburning stoves are becoming increasingly trendy in wealthy households in the U.K. The misguided belief here is that burning wood, now rebranded as biomass, will help to alleviate global warming and save carbon emissions. The idea that burning wood is somehow carbon neutral is plainly ludicrous. More than one million British households now have a wood burning stove and more are being installed each year to the detriment of local air quality.

It seems that our Clean Air Act which saved the lives of many Londoners back in the 1950`s is now becoming all but redundant. Back in the 1950`s open fires were very common since central heating did not exist in many homes. This caused the skies of London and other cities to become filled with dark acrid smoke. During foggy periods the skies of London became known as foggy pea soupers. These smogs gave rise to a lot of serious ill health and respiratory illnesses became rife.

The crippling air pollution forced Governments to act and so a Clean Air Act was introduced in 1956.London skies were once filled with smogs from countless open coal or wood fires and chimneys belched out noxious smoke. Not surprisingly respiratory disease was rife and tuberculosis was very common a century ago up until the 1950`s.The killer smog of London in 1952 was mainly caused by these countless open fires. The terrible toll on public health finally gave rise to the Clean Air Act of 1956.Another Clean Air Act was introduced in 1993.

However these vital protections to our health are now being reversed by Whitehall "climate" legislation. DEFRA has called for rules to be relaxed as the Clean Air Act now conflicts with low carbon energy policy! This new policy means that it is now encouraged to burn wood in fireplaces as it is now supposedly low carbon. In fact wood is not at all low carbon and burning wood will cause serious damage to the health of all local residents. Neither is it acceptable to burn coal in the fireplace for the same reason.

I vividly recall a green energy open day that was held in my street to demonstrate eco energy. It was sunny day with clear blue skies in October. As I walked past the eco open day at the house I saw a jet-black plume of smoke belching out of the chimney that looked and smelt awful. Inside the house were two new woodburners blazing away at full blast with logs inside. This event many years ago got me thinking about burning wood or biomass.

Although it looked very cosy and homely inside, the outside of the street was fast becoming a polluted nightmare and I prayed the eco day would soon end as the acrid choking smoke blew into my kitchen. I appreciate that open fires are lovely to look at. For isolated rural dwellers surrounded by woods, I see no harm in their burning log fires.

However there can be no questioning that log burners will prove useful in America for the self sufficient preppers. If an ice age arrives, those with a supply of wood to burn and a log burner may fare better than those reliant on the national grid. Survivalists who have ample land and who live in secluded woods can easily afford to burn the fallen wood and will not jeopardise the health of residents. As a last resort, burning wood may be the only means to cook food and to keep warm.

The preppers in America are ready to live off the land and will be able to survive when an apocalypse arrives. When the national grid collapses then clearly people will need to resort to burning wood. When power cuts arrive, burning home gathered wood will help the rural survivalists to survive an ice age. Burning wood to generate heat is fine in isolated areas or countries with vast tracts of available land. However many studies are now showing just how dangerous burning wood in congested built up areas can be. Wood smoke releases some benzene compounds that are linked with childhood leukaemia. Benzene is a carcinogen that is also found in petrol emissions.

This trend for burning wood for energy is leading to widespread deforestation worldwide. Some of our power stations here in Great Britain have been converted to burn biomass. This is probably not a great environmental policy as the world needs more trees. Furthermore burning wood releases a lot of noxious chemicals linked with cancer, bronchitis and heart disease. Burning wood in a densely populated residential area will lead to respiratory problems and particulate pollution in the neighbourhood. The idea that it is somehow "greener" to burn wood is a contentious one.

A study published by New Scientist showed that inhaling wood smoke for an hour is equal to smoking 100 cigarettes. It is vital that the wood to be burnt is carefully treated and dried but this rarely happens. Burning any old wood will not do and may release deadly dioxins, heavy metals and pesticides.

Sadly many trees are now being destroyed for wood-burners and not being replanted. It seems to be a case of not being able to see the wood for the trees! Burning trees for fuel is not eco-friendly, as trees give out oxygen and take up carbon dioxide. Unlike the inert coal used in power stations, trees are living with leaves. We have precious little ancient woodland left in our country.

Shotover Country Park is a nature reserve that I often visit and is owned by Oxford City Council. The park rangers have been felling scores of ancient majestic trees and turning them into woodchips for trendy wood chip boilers. I believe the council is slowly destroying this nature reserve. The council is not replanting any of the mature trees it has felled for firewood, so the idea of wood being a carbon neutral fuel is, frankly, a joke.

Ironically in our bid to be "green" we are now burning more wood than at any time since the Industrial Revolution. The reason we are turning to wood is that we will lose a lot of energy when all of our 15 coal power stations close to conform to an EU directive. This Directive is known as the European Large Combustion Plant Directive and was instigated by Euro Green MEP`s who hoped that closing coal power stations would stave off an imminent global warming. This premature closure of fully functioning power stations not at the end of their shelf life will likely lead to an energy gap and energy insecurity.

The sad aspect of our power station closures is that it is entirely unnecessary. The coal power stations in the U.K benefited from advanced technology to clean up emissions. Their chimney stacks had scrubbers that removed most of the particulate pollution that may cause acid rain. There was no harmful pollution around the coal power stations here in the U.K. due to technologically advanced chimney scrubbers that scrubbed the particulate sulphur pollution from the emissions.

I witnessed the Power Station at Didcot, Oxfordshire being blown up. Everyone was cheering as the large towers collapsed but I was aghast. Many years before this futile and pointless destruction the Oxford Mail and Oxford Times had published several of my letters the gist of which was that we should keep Didcot Power station and our other coal power stations. When Didcot power station in Oxfordshire was up and running did it lead to an ambient rise in local temperatures as the warming coal emissions model would predict?

Actually the converse was true. There was a measured statistical increase of a chance of winter snow flurries around the power station! So it seems that the model of carbon dioxide as a warming gas may have been overstated and that we have more to fear from a global cooling caused by our emissions. Did the health of Didcot residents suffer from living near the power plant?

No, it most certainly did not suffer from the clean emissions. Didcot had benefited from advanced technology with chimney stack scrubbers removing all particulate pollution though the carbon dioxide remained. Epidemiologists in Oxfordshire found that residents of Didcot were not only healthier than average but also had a slightly longer life expectancy! The only emissions from our cheap to run power stations were steam and a smidgeon of carbon dioxide.

Instead of closing our vital infrastructure, we could have planted a small copse of trees around each power station. This would have satisfied the need to offset carbon emissions and pleased environmentalists who are concerned about carbon dioxide emissions. Trees also soak up carbon dioxide lie a thirsty sponge. I wrote to the Government with this suggestion to plant a copse of trees around each power station ear marked for closure to absorb the carbon dioxide from the power stations.

This pragmatic suggestion has not been implemented needless to say! The Government would rather run the risk of power cuts and impose expensive energy from nuclear power on its citizens instead of planting some trees. The closure of coal power stations seems to have become a symbol of our efforts to care for the environment. It is mere gesture politics.

The closure of our desperately needed power stations will lead to power cuts if we do not replace the energy deficit. There are far better ways to look after our precious ecosystems. People can try and cycle instead of driving their car. The worst emissions are undoubtedly those from diesel vehicles that emit particulates and nitrogen dioxide which harm our health.

Blowing up perfectly good power stations to save the planet is the most ludicrous idea that any civilised country has embraced. While it is true that burning raw coal in open fires releases unpleasant pollutants, this is not the case when coal is burnt in our advanced power stations. The irony is that by closing our power stations our skies may become far more polluted and lead to global dimming.

When the power cuts arrive in the future homeowners will increasingly turn to burning wood or untreated coal in their fireplaces to keep warm! Our advanced power station technology to burn coal cleanly for electricity will be replaced with polluting wood stoves and open fires. We are indeed regressing back into the Dark Ages! If the U.K continues its love affair with burning wood the resulting hazy pollution may create an added risk for dimming the sun`s input. This may further exacerbate any dip in temperatures and encourage the onset of our well overdue ice age.

Many studies have proved that air pollution actually prevents sunlight from reaching us and therefore smogs cool the ambient temperature rather than raise the temperature. A Channel 4 newscaster recently announced that smog would lead to more global warming and this assertion was incorrect. This cooling effect from aerosols has been demonstrated repeatedly by volcanic eruptions that emit airborne particulate aerosol matter.

The volcanic eruptions of Mount Pinatubo and Tambora resulted in lower temperatures worldwide that are popularly known as the years without a summer. Smogs and dense particulate pollution may harm people, but should not be confused with pure carbon dioxide emissions. Carbon dioxide as a pure gas is both good for plants and ocean life.

Air pollution has repeatedly been shown to lower rather than raise temperatures. For example after the 9/11 air attacks all planes were grounded in New York. Because the skies were clear of hazy aircraft contrails, the sunlight was able to reach the ground unfiltered. The measured temperatures shot up in New York. This meteorological observation demonstrated that the aircraft contrails are cooling temperatures over cities such as New York. When the ice age arrives, many thriving American cities such as New York will be hit very hard.

Back in 1974 Wilson Clark wrote that "instead of growing warmer, the Earth may enter an ice age as a result of fuel combustion. The combustion of fossil fuels releases large quantities of particulate matter into the atmosphere, which may reflect sunlight away from the Earth, thus cooling the planet". So here we have a climate hypothesis that is diametrically opposed to the present day theory that fossil fuels may cause global warming.

Instead Clark states that these fossil fuel emissions may cause a drop in ambient temperatures and that our industrial activity may cool rather than warm our atmosphere. This seems a far more persuasive argument than the warming hypothesis .It is interesting to note that just as today there is a scare about global warming , back in the 1970`s there was a similar scare about a global cooling and a coming ice age. The global warming fear has probably become the dominant one since it now has huge financial vested commercial interests backing it.

Either way we should be cleaning our skies. With the amount of air, ship and vehicular traffic, this looks nigh on an impossible task. The last thing we should be doing to keep our skies clean in the U.K is to blow up perfectly good power stations. We need our infrastructure to keep our society and economy functioning. There are far better ways to ensure our skies are clear. Households should avoid installing a trendy woodburner and avoid burning any fossil fuel such as wood or coal in their fireplace.

Nitrogen Dioxide (NO_2) is a common pollutant in towns and cities, and diesel vehicles are a particular problem. According to think tank Policy Exchange the most recent Euro 6 diesel cars emit more than five times as much NOx as Euro 6 petrol cars. Euro 6 are the latest EU standards for air pollution emissions from cars (European Commission 27/06/2016).

Yet diesel cars were promoted as being good for the environment as they are allegedly low in carbon emissions. This is just one small example of the many scams that have arisen as a result of the carbon dioxide legislation. Electric cars are the way forward as they emit no particulate pollution. Therefore thousands of lives will be saved. However when the ice age arrives we may need 4 wheel drive versions with extra gripping tyres! Worryingly we might not have enough electricity to charge up the electric vehicles. As the climate cools there may be an energy deficit.

This legislation to promote “low carbon” diesel cars came from the EU and it seems that the EU is causing more harm than good with some of its Directives. Despite this legislation for clean emissions there are around 40,000 premature deaths a year now in the U.K. thought to be caused by air pollution. Far from improving matters the legislation introduced to help prevent global warming is making things worse. This is just one of the examples of how the focus on carbon emissions has resulted in unwise legislation.

The other example that springs to mind is the closure of our non polluting coal power stations in the U.K that did not emit any harmful pollutant. Another example is the fashion for laughably "low carbon" woodburners that are now causing air pollution levels in London not seen since the days of the foggy pea soupers back in the 1950`s . All done in the name of preventing a man made global warming event.

So it seems that some of misguided attempts to lower carbon emissions have resulted in greater air pollution in our cities. DEFRA has called for air quality rules to be relaxed as the Clean Air Act now conflicts with their low carbon energy policy. This new policy means that it is now encouraged to burn wood in fireplaces as it is now supposedly low carbon! The policy to burn wood was probably introduced to make up for a deficit in energy caused by closing our coal power stations.

Clearer non-polluted skies will lead to greater sunlight radiation reaching Earth and warmer temperatures will boost crop production. Perhaps we should curtail burning fossil fuels to prevent a global cooling rather than a global warming. Whichever climate theory you may subscribe to, it should be imperative that we do not pollute our skies with smogs.

NASA has recently confirmed that the Northern Hemisphere may get cooler as a result of sulphur aerosols and pollution. The Goddard Institute of Space Studies has published a paper by Gavin Schmidt and Kate Marvel showing how aerosols cool temperatures. This theory of cooling aerosols was discussed in the early 1990`s by climate scientist Antony Milne in his excellent book "Earth`s Changing Climate" which I highly recommend.

NASA has only recently decided to include pollution in their climate models. It looks as if their conclusion that burning fossil fuels may cool the Northern Hemisphere might be the most accurate deduction of their climate modelling. NASA says that carbon dioxide is distributed evenly around the globe to cause a uniform warming effect. That is factually incorrect even according to NASA`s own satellites!! The carbon dioxide does not mix uniformly in the air and instead gathers in pockets before falling to the ground as it is a heavy gas.

NASA has satellites photos showing this patchiness of carbon dioxide around the globe. NASA cannot therefore state that carbon dioxide is distributed evenly in the atmosphere!! The satellites pictures show as clear as daylight that the carbon dioxide is not evenly distributed at all and in many parts of the atmosphere it is completely absent! This one single fact alone punches a huge hole in the carbon dioxide warming hypothesis.

Kate Marvel who is the lead author of the cooling aerosol study writes that the aerosols are more or less confined to the Northern Hemisphere, where most of us live and emit pollution. There is proportionately more land in the Northern Hemisphere than the Southern Hemisphere. Land reacts quicker than the ocean does to these atmospheric changes (Marvel, Schmidt et al 2016).

One of the reasons that burning fossil fuels such as coal can cause cooling is that it will release sulphur dioxide. However when treated fossil fuels are burnt in a coal power station the sulphur dioxide will be removed by chimney scrubbers and this problem will be avoided. China however is burning a lot of untreated coal. NASA has said that China is one of the most polluted countries in the world and the smog can be seen from space satellites. Clean transparent air allows sunlight to reach us warming our skin and improving our mood.

Imagine the dawn of the new ice age when the sun goes dim and skies are leaden and grey. This is a real prospect for us in the not too distant future. Meanwhile some climate scientists are convinced that there is a need to engineer some sort of a cooling sunshade over the Earth to prevent global warming! Some mad -cap scientists are now exploring ways to further cool the planet! This topic of geo-engineering the climate will now be discussed.

Chapter 10: The Hazards of Geo-engineering our Climate

Scientists have long known how to seed clouds with silver iodide in order to make it rain heavily. In 1946 a scientist named Irving Langmuir working for General Electric discovered how to make rain in the laboratory. Silver iodide was found to be a good nucleating agent for the seeding of rain clouds. The science of geoengineering the weather had arrived.

When it comes to tampering with our weather one should adopt a precautionary approach. A cloud seeding experiment was once conducted in Devon using silver iodide as the cloud seeding agent. This military experiment resulted in torrential rains leading to flash floods, mudslides and deaths. On August 15, 1952, one of the worst flash floods ever to have occurred in Britain swept through the Devon village of Lynmouth.

Thirty five people died as a torrent of 90m tons of water and thousands of tons of rock poured off saturated Exmoor and into the village destroying homes, bridges, shops and hotels. New evidence from previously classified government files suggests that a team of international scientists working with the RAF was experimenting with artificial rainmaking in southern Britain in the same week and could be implicated. Certainly this was a suspicious deluge of rain that appeared to come from nowhere.

Later an American scientist admitted taking part in Project Cumulus to make it rain hard in Britain at that time. Today it is reported that cloud seeding takes place over ski resorts to ensure good snow cover. Japan has used seeding for extra rain to enhance crop production. Colorado ski resorts often use this method to boost snow cover. The Wyoming Weather Modification Pilot Project indicates that snow fall can be increased by up to 15 per cent using this method.

So it seems that man can indeed influence the weather and therefore the anthroposphere is a real concept. Recently scientists have been turning their attention to ambitious geo-engineering projects to cool our entire planet! No doubt the residents of Norway and freezing Russia will be most appreciative! Since it has been shown from geologic records that a very small change of insolation can dramatically cool our planet, this gigantic sun canopy could trigger a massive ice age.

I once attended a public climate change lecture in the Oxford Physics Department where scientists were discussing various ideas of how to cool the planet with carbon capture and geo-engineering. There was a futuristic and mind boggling proposal to engineer a giant sun canopy shield over our atmosphere to keep our planet cool! The main "solution" to cool our world is to inject reflective sulphur based aerosol particles high up into the atmosphere. These particles will reflect the sun`s light and heat back into space , in much the same way that volcanic ash cools the skies as discussed in the previous chapter.

The National Center for Atmospheric Research suggested in 2006 that sulphur aerosols should be injected into the atmosphere to keep the temperature down. Tom Wigley proposed that the amount of sulphate aerosols should be of a similar quantity to those aerosols ejected by the volcanic eruption of Mount Pinatubo in 1991 which cause temperatures to fall in the region. When the volcano erupted in 1991 the following year temperatures fell.

This is clearly a vast amount of airborne matter to be put into our stratosphere. Tom Wigley thinks that this geo-engineering project will be able to "buy us time" in which to deal with the challenges of global warming. Wigley convinced himself that geo-engineering the atmosphere would prevent a carbon dioxide induced temperature rise. Or maybe he just wanted to make a profit! This worrying project shows just how far scientists are willing to go in order to combat the perceived anthropogenic warming.

It is known that huge amounts of sulphur based aerosol will lead to a lot of damage to our precious ozone layer. We are fortunate that Wigley`s proposal to mimic the output of a yearly volcanic eruption did not come to fruition. It is thought that one of the reasons for the cool temperatures of the little Ice Age could have been due to five volcanic eruptions during that time.

The respected scientist Nigel Calder has voiced his cynicism over the alleged warming properties of carbon dioxide. Calder proposes that the greenhouse effect of carbon dioxide levels off after a certain point of saturation has been reached. So rather than a hockey stick we have a level graph. In other words more carbon dioxide does not equate to more warming. It is as simple as that. Therefore, according to the late Nigel Calder, most, if not all, of the accepted models of warming by carbon dioxide are incorrect.

This assertion is given further weight by the observation that although carbon dioxide levels continue to rise to the present day level of around 400 parts per million the observed global temperatures are not rising in tandem with the measured CO_2. In other words although carbon dioxide is a valuable greenhouse gas that has a benign influence its effects are limited. Therefore the whole industry that has sprung up around saving carbon emissions may turn out to have been a red herring.

However it has been a lucrative endeavour for those in the low carbon business loop! Meanwhile a whole generation of scientists are blindly going down a dead end in their quest to geo-engineer a cooler planet. The main problem facing our survival is the onset of a major ice age. By the time this overdue ice age has arrived it may be too late and whole swathes of the Northern Hemispheric population may be wiped out. Even worse is the realisation that to date all of our efforts to make our world safer have in fact made it a far more dangerous place.

So far all of the proposed ideas to geo-engineer our climate, revolve around the need to cool planet Earth. Not one proposal has emerged to geo-engineer our climate in the face of a bitter ice age. However the USA has a military project called HAARP which some conspiracy theorists believe is to heat the ionosphere in the advent of an ice age.

Some worried Americans are convinced that their Government knows something about a rapid cooling event and that this project is working on a plan to avert a catastrophe. This is speculation however as the HAARP (High Frequency Active Auroral Research Programme) project was designed to study radio transmission signals high in the ionosphere for a military purpose. HAARP did use an ionospheric heater as part of the research though it is unlikely it had anything to do with preparing for an ice age.

Another project named EISCAT involves an ionospheric heater and took place in Norway. HAARP is loosely based on this work. EISCAT stands for European Incoherent Scatter Scientific Association. Again the research involves radar signals high in the ionosphere. I am presuming the purpose of these studies is to understand the interaction of solar influences on radar. I do not fully understand how an ionospheric heater is being used as I am not an expert in this field.

It seems that the ionospheric heat pump studies plasma turbulence using high frequency transmitters that reflect radio waves back to the ground from the ionosphere. Whether this device can be used to heat the atmosphere is purely speculative. However it seems apparent that there are strange experiments taking place in the skies above us .The Oxford scientist Pak -Han Wong is advising that we adopt a precautionary and ethical approach to geo-engineering experiments.

Not one geo-engineering proposal to date involves warming the planet to prepare for an ice age. I have to admit that I was concerned on hearing of the proposed giant sunshades as in 2007 I had already considered the prospect of an impending ice age. There are many plans in the pipeline for carbon capture projects. However some scientists feel that carbon capture technology may prove prohibitively expensive. The Earth`s own carbon cycle already does a good job at capturing and recycling carbon. Carbon dioxide becomes carbonic acid when falling as rain and this in turn becomes calcium carbonate and so on.

I attended a climate conference in Oxford that discussed carbon capture and the engineers mostly agreed that it would be prohibitively expensive. At the post drinks reception I chatted to a few engineers and despite the free alcoholic drinks they seemed quite alarmed at the some of the plans to capture carbon dioxide from power plants. One of the engineers was very worried and said that the proposal to capture carbon would be ridiculously expensive and was not an economically viable engineering proposition!

Fast forward ten years to the year 2018 and the sensible warnings from engineers have been completely ignored as expensive carbon capture pipelines are planned for Drax power station. An example of a carbon capture and storage power plant is the Petra Nova plant in Houston. This large post combustion CCS (Carbon Capture and Storage) facility is costing billions of dollars!

The process of adding the carbon capture and storage doubles the cost of the power plant. It is not necessary to do this to avert global warming either. However some "good" may yet come out of the expensive upgrades at power plants. The carbon dioxide gas may be used to pump out crude oil it seems in a process called enhanced oil delivery.

Other geo-engineering proposals include dropping large amounts of iron filings into the oceans. The iron will encourage the growth of phytoplankton and lead to a green algal bloom. The plankton will then remove excess carbon dioxide and theoretically this will help to cool the planet. David Biello reports the controversial carbon sink iron experiment in the *Scientific American* (July 19, 2012.).

In 2012 a geo-engineering experiment took place in the ocean that was deemed to have violated United Nations rules and conventions on geo-engineering. An American businessman dumped 100 tons of iron sulphate into the Pacific Ocean on the West coast of Canada. This caused a huge algal bloom of plankton that could be seen from space satellites. The bloom covered an enormous area that was thousands of square miles large. The dumped iron had acted as a giant fertiliser for the plankton.

The reason that business man Russ George conducted this geo-engineering venture was that he hoped to make a fortune from the lucrative carbon credits market. This carbon trading business has been discussed elsewhere in this book. The carbon market has also served the skew the impartial debate on climate change in favour of global warming. As a financier friend of mine once said "always follow the money my dear!"

The theory is that the green plankton will remove carbon dioxide from the atmosphere and so the businessman would make a large profit with carbon credit schemes. However scientists are still debating whether this idea may be harmful to ocean life. There are concerns that experiments such as these could harm ocean eco systems and produce toxic tides of oxygen depleted waters.

The lucrative carbon credit scam was sold to the local community as being of environmental benefit. The villagers were told that the iron filings would enhance their salmon fishing. Legal experts said that the project had violated the UN convention on biological diversity. All of this tampering and attempts to cool our globe may have catastrophic consequences and lead us into a new ice age even sooner than anticipated. These ice ages are not trivial weather events such as an annoyingly cold winter, but cataclysmic events lasting up to 100, 000 years!

We have more to fear from a massive ice age and should therefore leave the geo-engineering to the proverbial mad cap eccentric scientists! There is a widespread belief that sinister and secretive geo-engineering experiments may already be taking place across the world. A more likely explanation is that the hazy contrails created by aircraft and ships are reflecting sunlight back and so are incidentally creating a geo-engineering cooling effect that is not intentional. This contrail hypothesis could be corroborated by the fact that when aircraft were grounded during the terrorist attack of 9/11, the local ambient temperatures rose.

Meteorologists measured a localised rise in temperature in New York as the lack of air craft contrails meant that sunlight could reach the ground surface unimpeded. As an aircraft contrail dissipates, it leaves behind an icy haze. The sky may appear cloud free, but the particles are there until they fall out of the atmosphere. The contrail particles scatter the sun's light in a similar way as in the proposed geoengineering cooling projects.

Therefore since our skies are already full of cooling particles it seems a redundant policy to geoengineer further light reflecting particles. Furthermore it is known that these cooling sulphur aerosols can damage the ozone layer. The ozone layer is vital since it shields us from incoming radiation. Therefore it is of concern to learn that in 2012 two Harvard engineers conducted a geoengineering experiment over New Mexico. The scientists used a hot air balloon to spray light reflecting sulphur aerosols that were similar to those cooling particulates emitted by volcanoes.

A similar geoengineering event here in the United Kingdom aroused controversy and was eventually cancelled. The U. K. project was known as SPICE which is the acronym for Stratospheric Particle

Injection for Climate Engineering. SPICE was put to rest after campaigners argued that it could open the door for a large scale implementation of the concept. The project was funded by a large Government grant and Dr. Matthew Watson was the study`s lead author. It seems that a dispute over patents and copyright may have been the final straw for the research.

However scientists are still looking at ways of cooling our planet. This is very worrying as the hard data from the geologic record is warning us that our Northern Hemisphere is heading for a major glaciation event. This event could arrive sooner than expected and we will not be prepared for it. Do we really need any extra cooling aerosols anyway when there are already so many emissions from volcanic activity and aircraft?

There has been a lot of internet chatter about an alleged Government conspiracy to affect our environment with chemtrails. These chemtrails sprays are thought to be sprayed from aircraft and supposedly look just like the emissions from aircrafts known as contrails. While researching the nefarious art of geoengineering techniques I did indeed come across experiments that might have given rise to this conspiracy theory. It seems that there are numerous ongoing studies into how to use aircraft to deliver aerosols high into the atmosphere in order to cool our planet.

Most of these aircraft spraying research projects are taking place in North America which is rather ironic as America has recently suffered from some appallingly cold snowy winters. The press reports show footage of the homeless sitting on top of heating air vents in the streets in Washington DC and struggling to survive. Yet America is home to the most sophisticated research being conducted in how to cool our climate.

A geoengineering paper on cooling aircraft spray was published in 2010 by Jeffry Pierce titled "Efficient formation of stratospheric aerosol for climate engineering by emission of condensable vapour from aircraft". So it seems that there is a grain of truth in the chemtrail conspiracy theory after all! As the old saying goes "there is no smoke without fire".

Residents in the Mount Shasta county of California are concerned about alleged geoengineering experiments using aluminium. They believe that aluminium is being sprayed using aircraft and that this toxic metal is getting into their water supply. They are convinced that aluminium products are being used in an experiment to cool the skies. It is possible that ski companies are behind this spraying if indeed it is taking place. It is a controversial area and there is no agreement yet, but a lingering suspicion remains over the cause of high levels of aluminium salts found on Mount Shasta.

Cooling and polluting emissions are also emitted by cargo ships according to a NASA study. Sulphur based aerosols have been linked with damage to the ozone layer. If the protective ozone layer is damaged it leads to greater radiation levels reaching us. This can damage our health and lead to genetic mutations. Some scientists even speculate that this radiation can lead to a higher risk of ailments such as thrombosis.

Any damage to the ozone layer can cause a hole in the ozone layer. These ozone holes let heat escape and this leads to global cooling. A team of Antarctic scientists think that the giant ozone hole is helping to keep Earth cool and as the hole mends the temperature may climb. Professor John Turner works for the British Antarctic survey and thinks that the polar ice increase may be due to the ozone hole above the South Pole.

The Montreal Protocol in 1989 banned ozone depleting chemicals such as chlorofluorocarbons (CFC`s) worldwide. Scientists hope that the ozone hole may heal by 2050 as a result of this ban. These CFC`s used to be found in spray cans but have now been banned from products such as hairsprays. Either way this ozone hole could be a warning sign of cooling and an ice age.

There are many substances that can harm the ozone layer which is found in the stratosphere. Geoengineering projects are also aimed at injecting particles into this delicate area. The appearance of the noctilucent polar stratospheric clouds may also be warning us that all is not well in the ozone layer. In a nutshell the signs are pointing to a dramatic cooling of our stratosphere. I believe that this could be a warning sign that we should start to prepare for an ice age of some description.

Everyone has heard of the term "nuclear winter". The worst case scenario of a nuclear war would be blackened skies that prevent all sunlight from reaching Earth. Even if the nuclear blast did not kill you the resultant nuclear winter would ensure that all plant life would die and we would slowly freeze to death. The term nuclear winter was first coined by Richard P. Turco who made computer models in 1983 of the sooty debris thrown up by nuclear war. His models suggested a catastrophic cooling if many nuclear weapons are detonated.

The climate scientist Stephen Schneider called this smoky layer a smokeosphere and his research papers published in 1986 that show a global cooling effect would result .We now have to ask ourselves a question about the atmospheric effects of our nuclear emissions to date. An enormous number of test nuclear bombs have been detonated above the ground. Surely these explosions must be having some sort of effect on our delicate atmosphere?

Since it is known that radioactive fallout can damage the ozone layer it is not unreasonable to question whether our nuclear emissions may have contributed to damage to the stratospheric ozone layer.. The weakening magnetic field is also causing ozone holes to grow larger and these growing holes will ultimately lead to heat loss in our thermosphere.

In February 2017 the North Korean Leader Kim Jong-Un launched a missile to celebrate his Father`s birthday! Missiles are constantly being tested all over the globe by the nuclear powers and we still do not know if these detonations may result in incipient damage to our fragile ozone layer. We do know however that detonating too many nuclear weapons in one go will lead to a nuclear winter, crop depletion and a severe ice age in the Northern hemisphere. Our skies are becoming progressively dirtier and this can lead to a cooling effect via global dimming.

Oxford has its own geoengineering programme at the Martin School. On its website it states that its aims "to engage with society about the issues associated with geoengineering and conduct research into some of the proposed techniques. The programme does not advocate implementing geoengineering, but it does advocate conducting research into the social, ethical and technical aspects of geoengineering. This research must be conducted in a transparent and socially informed manner."

This is mildly reassuring. However the website also states that "geoengineering is the deliberate, large-scale intervention in the Earth's natural systems to address climate change. Although we believe that society's first priority should be to reduce global carbon emissions, in dealing with climate change it may be wise to consider geo-engineering the climate to reduce the harmful levels of carbon dioxide in the atmosphere."

If we are indeed heading into an ice age the last thing we should be considering is measures to remove carbon dioxide from our atmosphere. Indeed we may perhaps consider the benign though improbable possibility that perhaps some of these greenhouse gas emissions may be stalling our overdue ice age.

If this is the case then we should be exceedingly thankful that carbon dioxide levels are rising slightly to 400 ppm. However we need to remember that more carbon dioxide does not necessarily equate to more warmth. The atmospheric dynamics are far more complex than this simple equation!

Highly respected scientists such as the late Sir Fred Hoyle and Nigel Calder believed that water vapour is a far more important green house gas than carbon dioxide. The respected astronomer demonstrated that as carbon dioxide levels rise, the greenhouse effect starts to diminish in a linear fashion.

So in other words there is an optimum level of carbon dioxide which serves to trap the radiative heat of the sun. After this optimum level of CO_2 is reached any rise in carbon dioxide does not correspond with an analogous rise in the greenhouse effect. Therefore any rises in carbon dioxide above this level will not lead to an exponential rise in temperature from the greenhouse effect.

This deduction is based on atmospheric physics and if it is correct then the logical conclusion is that even if our carbon dioxide levels are rising, as they seem to be at present, this rise will not necessarily lead to a similar rise in temperatures. The staggering conclusion from this assertion is that the linear models of rising carbon dioxide levels and the equivalent linear rises in temperature are probably deeply flawed.

None of this would probably matter too much were it not for the fact that we are now over due a major ice age. Yet all of our Governmental policies seem to be geared towards a warming event. It is important to remember that an ice age will affect the Northern hemisphere disproportionately as this is where the main land mass lies. This is why I gave the book the title “Ice Age Britain”.

Therefore equatorial regions will still enjoy warm and even hot weather as may have happened during previous glacials. The fact that equatorial regions may remain warm does not mean we should ignore our own very cold weather in the North. We are placing ourselves in a very dangerous situation by ignoring the coming glacial threat to our shores. The Northern Hemisphere is densely populated and we need to prepare ourselves for the fact that the predicted global warming may transpire to be a cooling event that triggers an ice age.

The main greenhouse gas is undoubtedly water vapour. The reason that we hardly ever hear of this fact is probably because there are no lucrative worldwide water vapour markets. No one is talking about curbing our water vapour emissions because frankly the movers and business shakers of this world have not yet figured out a way to make money from water vapour emissions! So all of the attention is diverted to carbon dioxide as this is where the big bucks lie.

The present day levels of carbon dioxide are not even that high in a paleoclimatic context. I can see some readers throwing their arms in the air aghast at this statement. However our present levels of atmospheric carbon dioxide are still substantially lower than in previous geologic eras such as the Ordovician- Silurian epoch when carbon dioxide was 4000ppm.

During the Jurassic-Cretaceous epoch carbon dioxide levels were at 2000ppm. Interestingly there was glaciation during the Ordovician epoch despite very high levels of carbon dioxide. However this is a very complex issue that would take an entire book to discuss and explore. One of the reasons for such discrepancies in CO_2 levels and climate in our paleoclimatic record is to do with rock formation.

When the tectonic plates collided to form the Tibetan plateau and Himalayas the very high mountain range caused a drop in carbon dioxide levels as rocks weather and remove CO_2. Therefore levels fell after this mountain range rose up. The fluctuations of carbon dioxide can be affected greatly by such carbon sinks as rocks and soil. If mankind is genuinely concerned about carbon dioxide levels they would curtail building and concreting over soil which is part of the carbon cycle. However the main problem is undoubtedly pollution and not carbon dioxide.

Pollution will be our undoing as our seas become choked with plastic microbeads from our facial scrubs and cosmetic products. These days up to one third of caught fish now contain minute plastic microbeads that have found their way into the oceans down our drains. A healthy food source of omega essential fatty acids is now becoming clogged with toxic plastic.

This pollution of our planet is where our attention should be directed. We are slowly poisoning ourselves with toxins that are now routinely found in food samples. Livestock are pumped full of antibiotics and crops are sprayed with poisons. Yet attention has been diverted to global warming, as this is where the big money lies!

A problem with the carbon dioxide warming models is that while carbon dioxide may help to warm the troposphere it seems to have the opposite effect in the stratosphere. The stratosphere is a layer of the atmosphere that is 9 to 14 miles high above the Earth. The stratosphere is also where our protective ozone layer resides.

High up in the stratosphere carbon dioxide actually acts as a cooling agent. So carbon dioxide works in opposite ways in the stratosphere and the troposphere. We are all familiar with dry ice which is a freezing form of carbon dioxide. It is possible that carbon dioxide has been overstated as a warming agent by the computer models. However when it comes to climate studies some of the best and faultless data comes from the geologic record.

There can be no disputing what has already occurred on our planet! This geologic record shows that for the past 2 million years there have been major ice ages lasting around 100, 000 years punctuated by interglacials lasting approximately 10, 000 years. We are now at the 11, 750 year mark in our interglacial, the Holocene, and hence our major ice age is possibly overdue.

The Mayan Long Count calendar predicted an end of a planetary era in 2012. Although I tend to be driven by hard logic and facts, a thought fleetingly crossed my mind that this major event might be referring to a new ice age. Such an event would most certainly mark the end of an old era and the beginning of a new epoch as the Long Count predicted. Intuition tells me that we are at the dawn of a new age, and this new epoch is the expected ice age!

As the late Sir Fred Hoyle rightly pointed out, the main reason our atmosphere traps heat is water vapour and the hydrosphere. If the water vapour were to turn into minute ice crystals over the Northern Hemisphere we would have an instant ice age as the shiny ice crystals would exert a strong light reflecting albedo effect. Water vapour makes a wonderful greenhouse gas until it turns to ice!

Yet scientists are still researching ways to geoengineer and cool our planet. A very minute layer of reflective particles around our atmosphere could precipitate an instant ice age as the late Sir Fred Hoyle wrote in his book called "Ice". In this book written in 1979 he describes how a very thin dust veil around the globe made from iced water vapour crystals or space dust would take a mere six months to initiate a full blown ice age!

Hence one can see just how very dangerous the proposal to geo-engineer our climate with minute aerosol particles would be. An irreversible ice age could start that would be self sustaining. Once the ice and snow of the ice age has arrived it would become self reinforcing because of the white ice reflective albedo effect.

As the Earth grows colder from the geo-engineered particles another feedback loop would kick in. As temperatures plummet the warming water vapour in our atmosphere would start to freeze into minute ice crystals. These sparkling ice crystals would shine like a transparent veil of diamond dust. The frozen crystals would then reflect even more of the sun`s radiation back into space thus beckoning a cataclysmic ice age.

Even if the geo-engineered particles were to eventually fall back to Earth the resultant ice crystals that formed would continue to sustain a most terrible ice age. It would be bitterly ironic if mankind succeeded in geo-engineering a new ice age when one is already arriving soon! It is possible that covert experiments are already taking place to cool our planet; in which case we can expect the mother of all ice ages!

Chapter 11: Global Warming; a Chinese Hoax?

The runaway popularity of the global warming theory has resulted in genuine perplexity amongst some politicians, the most notable of these being President Donald Trump. The American President became suspicious that China might be involved and he voiced his opinion that it might be a Chinese hoax!

Let us look at what is happening in China today. China has embraced an economic boom building huge tall concrete skyscrapers where once stunningly attractive pagodas stood. China has set about becoming a global economic super power by creating myriad manufacturing and construction jobs. Low level villages full of authentic Chinese charm have been bulldozed to make way for giant towering skyscrapers to maximise commercial profits.

In short China has set about changing its quaint past for a new concrete jungle of soulless skyscrapers. During the construction phase no doubt the Chinese economy soared as jobs were created. Now that the skyscrapers are finished they stand like grey sentinels watching over a desecrated landscape.

The massive building boom and industrial progress that fuelled an emergent tiger economy has inflicted an environmental problem on the citizens of China. Chinese city dwellers are often seen wearing face masks to protect their lungs from the terrible air pollution. The rush to modernise has been a disaster for China. Where once there were picturesque market streets with rickshaws there now are huge roads choked with cars. Even the skyscrapers seem to lie empty as no one wants to live in them or perhaps cannot even afford to.

The tiger economy may well start to fail soon as China appears to have thrown the baby out with bath water when it destroyed countless pretty small houses and historical villages. The smogs are a terrible legacy of this ill thought out expansionism. The result of this smog will be a decrease in natural daylight reaching the ground and China may well be contributing to a slight cooling effect. Not surprisingly China is now becoming more aware of sustainable and environmental issues. Recently China is embracing its rickshaw and cycling heritage and is encouraging city dwellers to take up cycling again.

China is experiencing dreadful air pollution as a result of the building bonanza. The expansion of the emergent tiger economy has come at a dreadful cost. Tall concrete skyscrapers have shot up everywhere creating a short term boom in the economy as builders were needed for the construction. Quaint Chinese villages were razed to the ground and much historic architecture was irrevocably lost. The authentic charm and beauty of Chinese architecture was speedily replaced with ugly grey concrete skyscrapers. Fools rush in where angels fear to tread! These gargantuan follies serve as a stark warning that a nation`s culture and history should be preserved and valued.

Far from benefitting the economy these monstrosities will in the long term damage the economy of China. This is because the landscape has now become very ugly and will no longer be attractive for tourism. Tourists wish to see the original and authentic China with its small welcoming markets and pagoda style roofs. Many homes of villagers were destroyed to make way for the skyscrapers that now stand empty. These gloomy sentinels serve a stark warning of the perils of unbridled building booms. Not only has China lost much of its inherent charm and history but a new problem has resulted from the building spree.

Where rickshaws once filled the narrow streets there is now a legacy of large streets filled with fast moving traffic. The terrible smogs in China are the result of vehicle exhaust particulate pollution. The change has taken place too quickly and the result is a terrible layer of smog hanging over the cities of China. Many rickshaws have been traded in for polluting vehicles as China rushes to become modernised.

As mentioned in the previous chapter, China has many new factories and coal power stations that are being built without chimney scrubbers. Therefore the skies have become seriously polluted. It is not unusual to see the hapless Chinese city dwellers wearing protective face masks as they struggle to breathe in the thick smogs. Certainly the rapid growth of the tiger economy has come at a terrible cost to the health of the Chinese who are now suffering respiratory disease in record numbers.

It is reported that over one million Chinese citizens now die from the air pollution. Many of the apartments in Beijing now have hermetically sealed windows and air purifiers. Understandably, the Chinese are now becoming very aware of environmental topics as the pollution exacts its toll on their well being. Therefore the Chinese are now desperately trying to undo some of the damage of their reckless building boom. Many cement factories were built to provide materials for the skyscrapers and these factories emitted pollution.

China`s new found enthusiasm for the environment is fully understandable as the Beijing smogs wreak a toll on health and the dingy hazy skies appear leaden and grey. City dwellers are sometimes trapped indoors unable to venture outside because of the appalling air quality. Many Beijing apartments now have air filters to purify the indoor air. The main problem is the PM2.5 particulates that are ultrafine.

These particulates can travel deep into the lungs and embed into the circulatory system. Particulate pollution causes inflammation to the arteries and leads to increases in heart disease, dementia and stroke .It is essential therefore that China start to clean its vehicle and factory emissions. As an aside, similar particulate sulphur aerosols have been proposed for the geoengineering cooling experiments.

The smog has resulted in a new found interest for ecology. In 2018 China announced plans to plant trees totalling the area of Ireland around the capital Beijing. The Chinese People`s Liberation Army has been dispatched to the smoggy province of Hebei with their shovels. The soldiers will plant thousands of acres of trees to alleviate the polluted air surrounding Beijing.

The Chinese are now embracing a nurturing approach to protecting their environment, though it does seem to be a case of shutting the stable door after the proverbial horse has bolted! Chinese steel factories continue to emit pollution as well as upsetting the world markets for steel. The overcapacity in the steel and cement production is considered to be an economic ticking time bomb for China.

The Chinese communist government has bowed before public anger and has now vowed to make the skies blue again! In 2017 the Government announced plans to close all of its 103 coal power stations. These plants unlike those in the United Kingdom, emit a lot of hazy particulate pollution The Chinese Government therefore propose to invest more in wind and solar technology to make up for the energy deficit. This is why global warming is being used as propaganda to justify the policy. It is not a hoax however but a need to clean up the skies.

Much of the particulates are coming from the heavy traffic that has now replaced the rickshaws and the narrow streets of the tiny villages. Had China not razed so many small communities to the ground and left its history intact then this air conditioned nightmare would not have transpired. This new found passion to protect the skies has led President Trump to conclude that global warming must be a Chinese hoax!

It is certainly not a Chinese hoax but a well intentioned blue sky thinking attempt to clean up the murky grey skies of China. A Channel 4 news report incorrectly stated that the Chinese smogs were caused by carbon dioxide and by global warming. This is a very common mistake that is often made. It is the particulate pollution rather than carbon dioxide that is causing the problems with air quality.

Carbon dioxide on its own is a colourless and clear, non toxic gas as the respected Nigel Calder, Editor of New Scientist would say in exasperation. Calder could not understand why this gas had been designated as a pollutant by Government legislation. Nigel Calder was of the old school type of scientist that did not enjoy close links with big business. He had probably failed to grasp the huge commercial profits being made when carbon dioxide had been designated a pollutant!

He was baffled and frustrated at the way the science was becoming skewed towards the demonization of carbon dioxide. In vain this editor of the respected magazine New Scientist tried to reason with those who were insisting that this harmless gas was now a dangerous pollutant. It was all to no avail. Carbon dioxide is a harmless gas that is non toxic. Yet carbon dioxide has now been designated as a pollutant with legislative measures. As soon as carbon dioxide was designated in law as a pollutant it enabled savvy carbon traders to become very rich.

I wrote to the British Government explaining why they had made a strategic error with the new legislation that made carbon dioxide a pollutant. I wrote that this error could cost our country £billions in lost energy generating capacity as we closed our power stations one by one. All to no avail. Carbon dioxide was now officially classed as a pollutant along with asbestos and all sorts of genuine life threatening toxins!

America also passed similar legislation to classify carbon dioxide a pollutant. No doubt this legal framework explains why carbon trading is so big both here and in the United States. Now we have the perfect storm of confusion reigning! The public begin to fear carbon dioxide as a dangerous pollutant. The public now believe that British power stations were polluting the skies when in fact they were doing no such thing.

The clean treated emissions from our coal power stations are now classified as pollution. The actual emissions are completely non toxic and harmless to human health. All of the burnt coal residues and sooty particulate matter have been removed by the chimney stack scrubbers here in the U.K. This is why houses have been built near coal power stations and residents have not suffered from any ill effects.

My local power station was Didcot in Oxfordshire and I used to look at the iconic cooling towers when I went for a walk on a high hill outside of Oxford. You could see the small clouds of steam sometimes when the caught the light in a certain direction. Sometimes the clouds would appear to be a leaden grey colour.

The clouds were not greyish because they were polluted but as an artefact of the way the light was reflecting on them. This confused some campaigners and bystanders who thought that this grey colouring was evidence of pollution. In fact the now defunct Didcot Power Plant was amongst the cleanest non polluting stations in the world!

After Didcot was blown up to comply with the carbon dioxide pollutant legislation, a lot of annoying power cuts disrupted Oxford. This reckless policy to close our non polluting power stations has already cost us over 8 .5 GW (gigawatts) from our electricity generating capacity! Carbon taxes that are generated from such closures will ensure that a handful of investors will become mighty rich!

One day I bumped into Oliver Tickell and decided to voice my concern at this train of events that had resulted from our coal plant closure. I was aware that the concept was partly based on his father Crispin Tickell`s famous speech on fossil fuels written for Margaret Thatcher. The speech given by the Conservative Prime Minister Margaret Thatcher in 1989 to promote low carbon nuclear power over coal power continues to cast a powerful spell even to this day. We should remind ourselves that the shrewd and canny Chancellor Nigel Lawson ultimately decided to shelve the idea because of the huge costs involved.

However I soon had a change of heart about lecturing him on the closures as one cannot fail to be charmed by Oliver, who has the most wonderful speaking voice like honey, and mischievous tiger eyes. “Not to worry” I said “the idea to close the coal power stations might be correct after all as the sulphur dioxide emissions can cool the atmosphere and we are probably heading towards a an ice age ; so perhaps closing them was in fact a serendipity".

However it did later occur to me that our own coal power stations have most of the cooling sulphur emissions removed by the chimney scrubbers. Therefore closing them will not be of any use in preventing the ice age. Indeed our power stations have no effect what so ever on our atmosphere. The only effect they have is to generate useful power for electricity!

Before the global warming zeitgeist became popular, scientists were voicing concerns about the imminent ice age. However carbon dioxide levels have risen recently and some optimists hope that this rise to 400 parts per million may be stalling the expected overdue ice age. This is because it is hypothesised carbon dioxide levels need to be below 280 ppm to facilitate the ice age and they are presently at about 360ppm to 400ppm. However as the old statistical saying goes" correlations are not causes" and we need to consider this fact before jumping to conclusions.

While carbon dioxide levels may indeed rise during warmer periods this could be an artefact of the warmer conditions. In other words the warm weather may encourage plant life and biological processes such as plant decomposition that will lead to small rises in CO2. Geothermal activity also releases large amounts of carbon dioxide. The Climate Research Unit in East Anglia has long maintained that global temperatures are rising fast.

This infamous research department has been cleared of any wrongdoing in the light of the "Climategate" debacle. Nevertheless, the disquieting fact remains that an email surfaced suggesting that an attempt was made to "hide the decline" in measured global temperatures. I do not feel that it served any purpose to conduct a witch hunt against an academic department that was under pressure to produce accurate statistics on global temperatures. This is no mean feat in the light of the logistics of taking accurate readings from poorly positioned weather stations.

In the Ordovician epoch CO_2 levels were around 4000 parts per million which is ten times higher than today. Yet there were still glacial episodes during the Ordovician epoch in our paleoclimatic history. In the Jurassic- Cretaceous period the level of carbon dioxide was around 2000 ppm which is five times higher than today. Then levels started to fall and today they are still relatively low in the context of paleohistory. What might have caused these falling levels of carbon dioxide?

Perhaps a collision of Earth`s tectonic plates 50 million years ago caused large amounts carbon dioxide to be sucked out of the atmosphere. The Himalayan mountain range and Tibetan plateau rose up thousands of feet above sea level as a result of the collision between the Indian Plate and Eurasian Plate.

This geologic process began 50 million years ago and continues today. The huge expanse of rock used up carbon dioxide in the weathering process .As long as huge mountain ranges exist on planet Earth the temperature is likely to remain on the cooler side since mountains act as massive carbon sinks.

A study published online in *Live Science* found that the Alps were acting as a huge carbon sink and that the rate of rock weathering was twice as high as expected. Large amounts of carbon are constantly locked up in limestone and soil on the mountains as a result of the weathering process.

Without the benign greenhouse effect in our troposphere the Earth would be locked into a permanent ice age. The late Sir Fred Hoyle felt that we should be thankful for a greenhouse effect and we should be adding more greenhouse gases rather than curtailing them. Hoyle felt that Earth was in far more danger from a returning ice age than from global warming.

I met Sir David King, former Chief Scientific Adviser to the government, at the inauguration of the James Martin 21st Century School in Oxford in 2007. This talented and jocular scientist mused that perhaps our carbon emissions were stalling the onset of this ice age. We both discussed the fact that during our last ice age U.K cities such as Manchester would have been buried beneath a mile thick ice sheet. Not very nice weather conditions for the Mancunian football supporters of Old Trafford!

However we need to consider the dire possibility that these greenhouse carbon emissions will not prevent this overdue ice age after all. A recent study by the Potsdam institute for Climate Research in Berlin grabbed the media attention in a big way. The headlines were declaring that the industrial activity and carbon emissions of humans have managed to stall our ice age by at least 50, 000 years! Phew! It looked like we have dodged a bullet with our man made emissions from burning fossil fuels (Ganopolski et al 2016).

However there is an ongoing debate about cause and effect regarding carbon dioxide and global temperatures that still has not been fully resolved. There is the problem of 800 year time lag dilemma. This time lag refers to findings that carbon dioxide levels rose 800 years after temperatures warmed up in the past on our planet. Therefore it looks as if rising carbon dioxide levels is an artefact of the warmer weather rather than the primary instigator. Therefore while there is undoubtedly a link between CO2 and temperature this link does not imply causality.

Antarctic ice core samples show that more carbon dioxide was released into the atmosphere after the planet thawed out. The oceans may have released more CO2 as Earth became warmer from increased sunlight. The greatest carbon sink on our planet is also in the soil. As the earth warms up the soil microbes and micro-organisms become more active and in turn release more carbon dioxide.

So yes there is a link between carbon dioxide and warmer temperatures but it still does not imply causality. Many climate scientists now assert that as the Earth warmed up in past eras, the carbon dioxide levels rose and caused a feedback effect, whereby temperatures then continued to rise even further during an interglacial.

The probable explanation is that as Earth warmed more plants and trees grew, and more dark soil was exposed to the sun as snow melted. This would create more dark patches for absorption of sunlight, raising temperatures further. Fauna and flora would flourish in a warmer climate and so more carbon dioxide would be exhaled since all living creatures exhale carbon dioxide.

As mentioned previously the Potsdam Climate institute`s Andrey Ganopolski, optimistically believes that we may skip a whole glacial cycle because of rising greenhouse gas emissions. According to the sanguine team of climate scientists, the extra 3 per cent of manmade carbon dioxide emissions will make the Holocene the longest and most stable period in climatic history! Did I see a flying pig go by!

We are now at a cycle of steadily decreasing solar insolation in the Northern Hemisphere. The solar radiation has been declining since the middle of our interglacial epoch, the Holocene climatic maximum which was around five thousand years ago. Lower sunlight or, reduced insolation is the main parameter needed for a new glacial inception.

However many scientists such as those at Potsdam and the Grantham Institute for Climate Change, are now convinced that we will avoid this ice age altogether because of our greenhouse gas emissions! However before breathing a sigh of relief that we have avoided a major glaciation event that might kill millions there is still a troubling issue to content with. The issue is the climate models of carbon dioxide and warming. The problem with climate modelling is that it is an inexact science.

Let us start with the simple fact that carbon dioxide in a test tube or bell jar does not behave in the same way as it does in our atmosphere. The atmosphere is a mixture of gases of which carbon dioxide is a very small proportion. Carbon dioxide is a heavy gas and tends to fall back down to Earth.

This is all part of the normal carbon cycle. The experiments conducted by the early pioneers such as Svante Arrhenius may well apply to a laboratory but really cannot be extrapolated to our complex atmosphere. The temperature changes of carbon dioxide observed in a Victorian bell jar cannot be applied to the whole atmosphere in an inference about climate.

Furthermore the experiments results are for the one gas only namely carbon dioxide. Our atmosphere is not comprised solely of carbon dioxide gas. Therefore the warming model of carbon dioxide taken from a bell jar may have been incorrectly applied to the whole atmosphere! Needless to say this statistical artefact or sleight of hand has produced some very alarming graphs about carbon dioxide induced warming. The temperature rises from carbon dioxide that we see in the global warming graphs are based on our entire atmosphere being made of carbon dioxide!

Furthermore carbon dioxide does not even mix well with other gases in our atmosphere. Instead it gathers in small pockets in the sky before falling back down to Earth where it will later become part of the White Cliffs of Dover which are made of calcium carbonate. Indeed most of the planet`s carbon is stored in the rocky lithosphere and the soil. Only a very small proportion of the total carbon is stored in coal. Carbon is mainly stored in the lithosphere as calcium carbonate and in the shells of marine animals.

A lot of the carbon sinks to the sea bed and some is later released in underwater volcanic emissions and so forth. The proportion of carbon dioxide released by us burning fossil fuels contributes about 3 per cent of the total carbon emissions in our atmosphere and the remaining 97 per cent comes from natural sources and processes. Now if say the situation was reversed and say that humans were emitting 97 per cent of emissions and natural sources were only 3 per cent, one might perhaps feel that the harmonious balance was being upset.

However it is really difficult to comprehend that a tiny three per cent extra carbon dioxide from burning fossil fuels is going to make a substantial difference to the 97 per cent that is emitted by natural processes into our atmosphere. Carbon dioxide is one of the weakest greenhouse gases. Our entire atmosphere is not comprised solely of carbon dioxide gas either. Therefore the model of carbon dioxide taken from a bell jar experiment may have been incorrectly applied to the whole atmosphere!

To reiterate this important point, any temperature changes that have been observed in a sample of carbon dioxide are applicable to the carbon dioxide sample only. These observed temperature fluctuations apply only to the sample of the carbon dioxide in the bell jar and not to the entire atmosphere, since our atmosphere is not made entirely of carbon dioxide. There is not a shred of a possibility that carbon dioxide could behave the same way in our vast atmosphere as it does in a test tube.

As soon as carbon dioxide rises high enough in our atmosphere it becomes chilly and starts to sink again. Carbon dioxide does not hang around in our atmosphere for long as most people have been led to believe. Carbon dioxide is constantly being removed from our atmosphere and becomes part of our carbon cycle here on Earth.

I realise that this may come as a drastic paradigm shift to those who have been persuaded that carbon dioxide is a dangerous gas that is building up in our atmosphere that will cause us to fry from a six degree rise in temperature. The hard facts are that this scenario from a carbon dioxide induced warming is nigh impossible. The basic laws of physics simply will not allow it!

There has been so much fuss being made about a carbon dioxide induced runaway global warming that even some very able scientists are starting to doubt themselves and their own inner convictions. Perhaps the Emperor does indeed have a nice new set of clothes and a kingly robe of ermine! This naked apparition of flesh that I see parading before the crowd is nothing but an illusion!

Since everyone is admiring these wonderful new clothes we too should join the crowd and praise this wonderful attire. Wanting to be part of the group or consensus is part of our human nature. Nobody likes to be left out of the group and everyone wants acceptance. If you then add financial incentives to the mix it could easily persuade some that the Emperor is indeed wearing a very fine set of new clothes indeed!

Recent experiments by Professor Nahle have demonstrated that any trapped back radiative heat transfer from carbon dioxide could only last for 5 milliseconds. This discovery is very important for the debate about carbon emissions and climate. His team`s finding plainly contradict all of the models that the IPCC have produced about carbon dioxide as a greenhouse gas. The scientists have demonstrated that far from being a fantastic warming gas, carbon dioxide loses most of its thermal energy instantly and cannot store heat.

The technical way of expressing this is that “the release of a quantum wave, at a different wavelength and frequency, lasts the time an excited electron takes to get back to its base state”. The hard scientific facts of the matter are that carbon dioxide is not much of a greenhouse gas at all. Therefore if we are pinning our hopes on this gas to stall an ice age we are in for a nasty shock! It looks like the Intergovernmental Panel on Climate Change may have made a few strategic errors in their climate model calculations.

The status quo must now be maintained as otherwise there would be an awful lot of red and embarrassed faces and this obduracy will doubtless save a lot of money in purchasing a new set of robes for the Emperor!! This is said without wishing to denigrate in any way the work of countless brilliant climate scientists who have toiled tirelessly over their graphs and models. The trouble with modelling is that the models are only as good as the data input and as the computer nerd saying goes garbage in garbage out or acronym GIGO.

The climate models all assume that temperatures rise in tandem with rising carbon dioxide levels in a linear fashion and that carbon dioxide behaves the same way in our skies as it does in a test tube. I believe that this is the main fallacy. As mentioned elsewhere in this book, the temperature does not continue to rise as carbon dioxide levels increase. There is a cut- off point or a saturation point after which carbon dioxide does not exert any greenhouse effect, if indeed it exerted much effect in the first place.

I recall a scientific adviser once saying that there is a little trick that can be used to demonstrate that carbon dioxide can heat the atmosphere. This trick takes place in a test tube or bell jar and is very convincing as a way to persuade others of a carbon dioxide induced warming. However as we now know, carbon dioxide does not behave in the same fashion in our complex moving atmosphere as it does when confined in a bell jar. Our planet`s atmosphere cannot really be compared to that of a bell jar and there is the carbon cycle to also consider. Our carbon cycle enables life to flourish on Earth as carbon is the main component of biological compounds.

It is rather ironic that President Donald Trump is blaming poor China for perpetrating what he describes as a "hoax of global warming" since it was the United States Supreme Court that designated carbon dioxide a pollutant! This legislation on carbon dioxide was then used to lower emissions from cars. As we know this legislation to supposedly lower carbon emissions led to various scandals whereby car manufacturers tried to cheat the emissions tests and were later fined.

No doubt the American legislation was partly inspired by Al Gore and his famous film about global warming called "An Inconvenient Truth". The truth of the matter is that carbon dioxide induced global warming is a genuine scientific theory that can be debated. However this scientific theory has been marketed, commercialised and advertised out of all proportion to its merits as a genuine scientific theory. Al Gore himself has made £billions from marketing alleged low carbon devices such as smart meters!

This has led to some genuine scientists without the commercial inks to feel genuinely aggrieved that the scientific debate of a theory is now being quashed. A scientist not in the carbon money loop is no longer able to query certain aspects the carbon dioxide global warming theory without being subjected to a ferocious attack or even called a "climate denier".

This expression "climate denier" has connotations with the abhorrent term "holocaust denier". Clearly to be a Holocaust denier is probably one of the worst things possible. Therefore this expression "climate change denier" was deliberately chosen by the vested interests. Indeed I have good reason to suspect that this awful term may have originated here in Oxford. It is a subtle form of psychological warfare.

Since nobody in their right minds would wish to be associated with being a Holocaust denier then it follows that no one will wish to be associated with the description of being a "climate change denier". A lot of covert psychological manipulation has taken place to ensure that ordinary, decent and principled folk who pride themselves on their integrity will be swayed towards the global warming theory. What decent citizen would ever wish to be placed in the evil camp of the "deniers"?

A formerly objective and impartial scientific debate has now become ugly and personal and emotional. Science should not be an emotional subject where feelings run high. Nowadays genuinely worried protesters often line the streets as they panic about a runaway "climate change". Facts are best debated in a cold and logical manner, similar to the unemotional temperament of the Vulcan Mr Spock from Star Trek!

Instead we have a highly charged overly emotional arena of debate revolving around climate change and climate change deniers. This is compounded with countless protests and actions by genuine environmentalists who have unwittingly been covertly manipulated by the carbon financiers and huge financial lobbies. The carbon profiteers even featured in the Paradise Papers taking advantage of dodgy tax loop holes.

Thus we have a perfect storm of climate change protests taking place as environmentalists and activists wish to make a stand against the perceived villains. Yet it is highly likely that some of these same big corporate vested interests are in fact pulling the strings and encouraging these demonstrations known as “climate actions”. The situation has become exceedingly complex and has many layers. No wonder then that some scientists are becoming frustrated.

A highly respected Professor of physics called Hal Lewis based at the University of California wrote a letter of resignation to the American physical society in 2010. The late professor Hal Lewis was of the increasingly rare old school breed of principled scientist who did not always follow the money .He resigned from the American Physical Society after 67 years of membership. Excerpts from his resignation letter are as follows.....

“It is of course, the global warming scam, with the trillions of dollars driving it and that has corrupted so many scientists, and has carried the APS (American Physical Society) along like a rogue wave. It is the greatest and most successful pseudoscientific fraud I have seen in my long life as a physicist. Anyone who has the faintest doubt should read the Climate Gate documents, which lay it bare.....So what has the APS, as an organization, done in the face of this challenge? It has accepted the corruption as the norm, and gone along with it... There are trillions of dollars involved, to say nothing of the fame and glory and frequent trips to exotic islands that go with being a member of the global warming club!”

The exotic islands he was referring to are the atoll islands of the Maldives. This low lying archipelago in the Indian Ocean has attracted much attention from scientists. Since the land mass of this attractive atoll is so low it is already uniquely vulnerable to flooding. Now the terrified islanders have to contend with the prospect of biblical storms and that sea levels will rise as global warming advances.

This deluged tropical paradise of 1200 islands has become the warning poster of climate change. The theory is that as the climate warms up the seas will rise and the Maldives will end up like the mythical world of Atlantis beneath the sea. Certainly the Maldives are vulnerable to flooding caused by tsunamis that may occur during a magnetic reversal.

Professor Hal Lewis was rather cynical however as he had seen too many researchers rush to the attractive Maldives to research global warming! It is interesting that he wrote this letter as Hal Lewis had himself once written a book on the possible technological risks of a man made global warming (Lewis 1992). So what had happened to make him attack a theory that he had once embraced? It seems that the principled scientist had become disenchanted at the huge amount of money being poured into an idea that he considered was still open to a scientific debate and there was not in his mind a firm scientific consensus.

The Professor was astutely aware that some of the data being used to back the global warming theory was invalid and had been manipulated. He felt that the financial lobbies had taken over the sacred realm of pure science and thus he resigned. Perhaps Hal Lewis should not have gone so far as to use the words “fraud and scam” since this undoubtedly alienated a lot of his colleagues .It might have been wiser to point out that there are scientific shortcomings in some of the carbon dioxide induced warming data sets instead.

The so called “Climategate” scandal was an example of a statistical mishap that took place in the U.K. at the University of East Anglia (UAE) in 2011 when around 5000 confidential e mails amongst research staff were hacked and released. This hacking resulted in a lot of embarrassing e-mails coming to light. The respected climate scientist Phil Jones was slightly implicated with a whiff of scandal. One of the e-mails speaks of how we “may try and hide the decline”.

This decline refers to an observed decline in global temperatures that had taken place since 1998. The general tone of the e-mails was one of concern that not all of the climate data was consistent with a global warming taking place on the Earth. The worried researchers may have been concerned that they had made errors in their data. This is fully understandable as this type of vast research is never going to be easy or straightforward.

An official enquiry was duly launched and the anxious Phil Jones was cleared of any wrong doing. Given the complexities of coming up with working models of climate that provide accurate predictions of temperatures this seems the only fair outcome. One of the worried e-mails that were leaked asks "what if climate change appears to be a multi decadal fluctuation? They`ll kill us probably!!..."The problem may have been that Phil Jones was very keen to please everyone and to come up with the expected data to show that global warming is increasing. When this data failed to emerge it must have aroused a panicked response from the baffled scientists.

However since this is a book about an ice age coming to our shores, I am rather pleased that the data from the University of East Anglia did indeed point to an overall decline in global temperature since the year 1998. This is exactly what one would expect the data to show if we are heading into the new ice age! Perhaps it would have been preferable if Phil Jones` team had looked into possible cause for the observed temperature decline such as pollution and solar variables. The team should not have discussed an attempt to disguise the data showing a decline in temperatures but instead tried to analyse the reasons for the observed decline.

A famous "hockey stick" graph was the result of an attempt to reconstruct the average Northern hemisphere temperature for the last one thousand years. The data used tree ring growth and other measures. It shows temperatures holding level for most of the 20th century and then shooting up like a hockey stick. This infamous graph has given rise to a heated debate. The physicist Richard Muller wrote that it is an artefact of poor mathematics in Technology review in 2004. However if it is indeed correctly produced, the graph does not invalidate the arrival of an ice age. Temperatures always shoot up just before an ice age!

The climate models that predict dire global warming exclude a lot of variables such as clouds and albedo and might be based on a flawed interpretation of correlation data. There is indeed some correlation in the ice core data between warmer climate and rising carbon dioxide levels. However the correlation is not always the cause. It is might be a misunderstanding based on flawed logic. The mistake may have arisen from a cause and effect process. When the weather warms up the microbes and organisms in the soil become increasingly active.

A recent article published by Michael Mann in the Scientific American magazine warns that by the year 2036 Earth will cross a threshold and be in the grip of a global warming nightmare. Michael Mann is also the co author of the very famous hockey stick climate graph. This controversial graph has been the yardstick of the IPCC (International Panel on Climate Change). The Intergovernmental Panel on Climate Change (IPCC) states in one of its predictions that carbon dioxide added to the air will raise the surface temperature by 1.4 watts per square meter. Where did they get this interesting statistic from I wonder??

Well it seems to bear a remarkable similarity to the 1.4 watts square meter statistic of heating of the Earth`s surface calculated by astrophysicists Henrik Svensmark and Nigel Marsh. Perhaps it has been conveniently borrowed? This amount of 1.4 watts has been calculated for surface heating when Earth`s cloud cover diminishes by 8 per cent. This total Earth cloud cover decreases during high solar activity. Quelle surprise!! An equation used for solar activity forced radiation seems to have been conveniently “borrowed” by the IPCC statisticians to measure a mythical carbon dioxide heating effect!

It is not surprising that many of us have been seduced by such numerical chicanery and artifice to convince us that we are in dire danger from carbon dioxide. After all not all of us are statisticians! When considering the climate it is important to realise that the portrayal of statistics can be misleading. For example some NASA data suggested that 2012 was the warmest year ever.

However a thorough examination of the data shows the average for global temperature was incorrect. Although the first eight months were indeed warm, the remaining months were far colder than average. Therefore the year 2012 overall was colder than average once the colder months were taken into account. It seems that you can play around with statistics to get the result you want!

I sense there is a question hanging in the air "is global warming a fraudulent scam or is it true that we are affecting the climate with our carbon dioxide emissions? The science of a greenhouse effect on Earth is factual and correct. There is a greenhouse effect that keeps the Earth cosy under a blanket of greenhouse gases. These gases include carbon dioxide. Therefore the greenhouse effect is real and not in any way a hoax. The greenhouse effect was originally a theory developed in 1896 by a chemist named Svante Arrhenius.

This nobel prize winning chemist was most definitely not Chinese and the alleged carbon dioxide hoax did not originate in China as Donald Trump suspects! The Swedish scientist Svante Arrhenius devised calculations to measure exactly how much of the Sun`s infrared heat was trapped by the carbon dioxide in the lower atmosphere.

Using the Stefan-Boltzmann equation Svante calculated that doubling the amount of carbon dioxide in the atmosphere would result in an increase in warmth on the ground. Note that he thought the rise in warmth would be at ground level only. Svante was in favour of a carbon dioxide induced rise in temperature due to the snowy climate of Sweden. The Swedish scientist therefore hoped that if enough wood fires were lit in Sweden then the temperature would rise as carbon dioxide emissions rose!

Svante had probably got the idea for his research from an earlier study by Irish scientist, John Tyndall. This idea of an envelope of warmth around the Earth was first put forward by John Tyndall who was born in 1820. Tyndall studied the capacities of various gases in the atmosphere to store radiant heat from the sun. Interestingly Tyndall found that water vapour was the greatest absorber of radiant heat in the atmosphere.

The other atmospheric gases lagged behind in heat absorption. These other gases with lesser heat absorption properties are carbon dioxide, nitrogen, oxygen, ozone, methane. Tyndall concluded that they exerted a very small heat absorption effect compared to water vapour Tyndall`s seminal and ground breaking experiments to measure heat absorption of various gases are were discussed in a recent book about the history of climate change ideas by Roger Fleming (Fleming 2005).

Tyndall found that the main greenhouse effect on Earth was exerted by water vapour. He gave a lecture in 1863 explaining how the Earth would be much colder at night were it not for the water vapour and other greenhouse gases keeping Earth warm. It is interesting to see how one of the minor greenhouse gases, namely carbon dioxide , has now shot to prominence as being the main greenhouse gas that is keeping Earth warm enough for life to survive. This seems to be the result of the huge financial lobbies now backing this theory.

Let us now return to the work of the chemist Svante Arrhenius who devised experiments to see if carbon dioxide was a potent greenhouse gas. He probably chose to focus on carbon dioxide as an industrial revolution was starting to unfold in Great Britain at that time .A lot of coal was being burnt in this era when coal was king. Great Britain was fast becoming the richest nation in the world as our coal reserves helped to drive the manufacturing industries that were springing up everywhere.

Arrhenius then predicted that our emissions from combustion of coal would be large enough to cause a warming effect. This was probably wishful thinking as he lived in cold snowy Sweden and was very hopeful for some global warming! He was probably the first scientist to make this observation and theory about anthropogenic warming. Interestingly another Swedish scientist named Knut Angstrom was very critical of Arrhenius.

Angstrom was a physicist born in 1857. He studied the absorption of our sun`s radiant heat by the Earth`s atmosphere. Angstrom felt that the radiant heat absorption of carbon dioxide was being over played. He believed that once a certain saturation point had been reached then adding more carbon dioxide would make no difference whatsoever to the greenhouse effect.

In other words adding more carbon dioxide does not equal more warmth. I find this fact interesting as it agrees with more recent statements by modern day scientists. Nigel Calder said exactly the same thing about the infrared saturation point of carbon dioxide in our atmosphere. So the climate graphs showing a meteoric rise in temperature as more carbon dioxide is added are in fact totally incorrect. To put it simply adding more carbon dioxide to our atmosphere does not make it warmer once the saturation point has been reached. Therefore all of the panic around rising carbon emissions is completely unfounded and based on a poor grasp of the atmospheric science.

The two Victorian scientists proceeded to exchange angry words and suffice to say that Svante Arrhenius was adamant that his warming theory of extra carbon dioxide was correct. Svante also hoped that the rising carbon dioxide levels from the industrial revolution would be sufficient to prevent another ice age from ever recurring on Earth. He was therefore all in favour of people burning as much wood and coal as possible!

Svante wrote "If the quantity of carbonic acid in the air should sink to one-half its present percentage, the temperature would fall by about 4°; a diminution to one-quarter would reduce the temperature by 8°. On the other hand, any doubling of the percentage of carbon dioxide in the air would raise the temperature of the earth's surface by 4 degrees; and if the carbon dioxide were increased fourfold, the temperature would rise by 8 degrees." (Arrhenius 1896).

This is the quasi science that is now being used in our modern times to promote a fear of carbon emissions. Arrhenius did not include feedback variables such as cloud cover in his models and so the theory was over simplified. The Swedish scientist felt that overall this hypothesised rise in temperatures would be a very good thing indeed as it would boost crop production and prevent another ice age. In those days any future global warming was considered as quite a propitious event!

However it seems that the entire carbon dioxide warming model was incorrect and that temperatures do not continue to rise as the level of carbon dioxide rises. There is a cut off saturation point beyond which adding extra green house gases makes no additional difference whatsoever to the climate. Since 97 per cent of all the carbon dioxide in our atmosphere comes from natural sources it seems highly unlikely that a contribution of 3 per cent by humans burning fossil fuels will make any difference.

Is it really feasible that our modest 3 per cent anthropogenic contribution from burning fossil fuels will make a difference to the 97 per cent that is emitted via natural processes? Most of the carbon dioxide is coming from natural sources such as volcanoes and forest fires. When animals and humans exhale they breathe out carbon dioxide also. Should we cull the animals and the human population to save carbon emissions?

We need to adopt a flexible approach and be prepared to embrace a paradigm shift that man is not causing any warming of the Earth through carbon emissions. If there is evidence for an anthropogenic effect on our climate, it is veering towards a cooling effect since pollution from dirty smoke cools the atmosphere. Specifically sulphur dioxide aerosols are proven to cool.

The evidence for carbon dioxide as a warming gas is very sketchy and there is compelling evidence pointing to it being a cooling agent in the stratosphere. The fossil fuel warming theory has recently been thrown into greater doubt after NASA found in 2016 that the Earth has cooled in areas of heavy industrialisation, where more trees have been lost and more fossil fuel burning takes place. So the exact opposite has taken place and cooling is the result of burning fossil fuels!

As more trees are cut down for biomass boilers the cooling will worsen in the Northern Hemisphere. The trees in Northern hemisphere forests also appear dark from space and so absorb heat. As more snow falls the albedo effect will also increase resulting in a loss of solar heat. As the cold weather starts to bite the United Kingdom may start to experience power shortages and power cuts. Well insulated homes will become mandatory.

The decision to close our coal power stations may be reversed as a matter of national security. Some crops may start to fail and more greenhouses will be needed. Some scientists are going so far as to say that our latitude may become a disabled zone for crops and agriculture leading to food shortages. The eighteenth century astronomer William Herschel in England noticed that the price of wheat was higher when sunspots were scarce.

President Trump is mistaken when he opines that global warming theory is a Chinese hoax since the theory of greenhouse gases has been around since the 1800`s. However since China has become so polluted China is now becoming more environmentally aware and keen to try technology to clean its skies. This desire for clean air is becoming confused with the ideology wrapped around carbon emissions and global warming. If there is an alleged hoax then it is far more prevalent in the USA and Great Britain than in China!

Al Gore has made popular films such as “An Inconvenient Truth” about anthropogenic global warming! Al Gore and many other hedge fund managers and American economists have made billions of dollars with carbon trading schemes and smart meters! It must have been rather a convenient truth for Al Gore that his film on global warming happened to help this business profiteering. The film “An Inconvenient Truth” is a documentary film highlighting the effect that humans are having on this planet and helped to energise environmental awareness.

The film was widely shown in schools and inspired the idea of sustainability. The blockbuster film was also rather convenient for Al Gore`s smart meter business interests and served to promote the profitable low carbon message. Al Gore confidently predicted that by the year 2008 all of the Arctic sea ice would have melted. He has now conveniently changed the forecast for the year 2050!! Interestingly this may be the same year that the next ice age arrives!

Chapter 12: Polar Vortex, Arctic Amplification Jet Streams.

We are all aware of just how unreliable our weather forecasts can be. I vividly recall the evening in 1987 when weather forecaster Michael Fish announced that a lady had phoned in to say that there was a hurricane on the way. He assured the television audience that there was no need to worry and that there was absolutely not a chance of a hurricane sweeping across the British Isles. Later that night I was woken by a ferocious howling noise of wind screeching through the gap in my window.

I peeked out through the curtains and noticed that a neighbouring silver birch tree that usually stood tall and straight was now completely bent at a right angle and the top was almost touching the ground! I had never seen a tree at such a strange angle before and it looked like it had bent in half. I then fell asleep and slept peacefully through one of the worst hurricanes to batter our shores. The nation awoke to scenes of chaos the following morning. This taught me a salutary lesson about the reliability weather forecasts!

Still there can be no doubt that the weather in the British Isles is greatly influenced by cold air masses such as the polar vortex and by the jet stream. The polar vortex and the jet stream work together in a synergistic fashion bringing weather systems to Great Britain. Therefore a discussion of the polar vortex and the jet stream is combined in this chapter.

The mass of cold air that hovers over our poles is known as a polar vortex. This term became a popular media buzzword during the recent freezing winters in North America and the U.K. When polar arctic air descends we have glacial conditions. Conversely when the milder westerly winds blow the weather is warmer and wetter during the winter months.

During our recent snowy winters, weather forecasters and meteorologists spoke of a freezing polar vortex that had become stuck for a lengthy period bringing exceptionally biting cold temperatures. There are actually two polar vortices. The stratospheric polar vortex is 19,800 metres (65,000 feet) above the surface of the Earth.

The tropospheric polar vortex is around 5,500 to 9,100 metres (18,000 to 30,000 feet) above the surface. Usually, when the weather forecasters are talking about the polar vortex, they're referring to the tropospheric vortex. Remember that the lower troposphere is the layer of the atmosphere where most of our weather patterns arise.

A polar vortex air mass is influenced by the position of the northern jet stream. The fast moving air flow known as the Jet stream can change direction from an East/West direction to a North/South direction over Great Britain. This enables the freezing polar air mass to descend over the British Isles. In the UK, extreme conditions were experienced in the winter of 2010.

The normal pattern of Arctic winds broke down causing a polar vortex to allow a frigid body of air to move south. This brought record snow conditions to northern Europe, eastern Asia and eastern North America and to Great Britain. According to the Office for National Statistics, there were nearly 44,000 excess winter deaths in the cold winter of 2014-15. A particularly cold winter puts elderly and vulnerable people at risk. There are heating problems as boilers breakdown and roads become dangerous.

The polar vortex descended over the British Isles in January 2017 bringing freezing conditions. Some scientists such as Timo Vihma from the Finnish Meterological Institute are hypothesising that the polar vortex is weakening over the Arctic due to a loss of sea ice. The question is being asked if the warming of the Arctic could cause the jet stream to meander and to wander south. When the polar vortex is strong it remains firmly in place over the North Pole.

When it weakens it descends bringing the polar air mass over the Northern hemisphere. However if melting arctic ice is the cause of the weakening polar vortex this would not explain why it does not always descend as in the mild winter of 2015. There may be other causes of aberrations in the polar vortex. The polar vortex may possibly be responding more to changes in the northerly jet stream instead of to melting Arctic ice.

Another theory proposes that the jet stream is becoming more undulating instead of remaining in a smooth tight circle above the Arctic. When the jet stream loops down it drags the cold air with it.

So the whole of Great Britain is then covered in a freezing polar air mass that normally would sit tight in the Arctic Circle. The jet stream is therefore becoming more kinked and wavy instead of remaining in a tight circular motion.

Jet streams are very fast moving air streams that are generated by the Coriolis force from the rotation of the Earth. The Coriolis force is caused by the rotation of earth on its axis. Jet streams are also affected by differences in temperature gradients from North to South. The jet stream in turn is affected by changes in solar activity. This is a very complex area worthy of research by an atmospheric physicist. Perhaps this phenomenon could be used as evidence that a global warming could lead to a cooling?

There is a phenomenon known as a sudden stratospheric warming which can affect the polar vortex in way that gives rise to colder temperatures for Great Britain. It may seem like a contradiction in terms to learn that a sudden warning of this very high stratospheric layer can result in cooling below. A Sudden Stratospheric Warming (SSW) event occurs when there is an abrupt warming of air, which is introduced into the cold Arctic polar vortex.

The reasons behind this are complex and the results are dramatic. An SSW event results when the cold arctic air mass gets pushed due south. The freezing cold air is carried south by the jet stream. An SSW cold weather event was responsible for freezing weeks of weather in Scotland in February 2018.This may seem confusing, so in a nutshell, a Sudden Stratospheric Warming event means freezing weather for Great Britain.

The stratospheric polar vortex can completely disintegrate and cause a wind flow pattern to reverse in the opposite direction. This can have an amplified effect on weather patterns in the troposphere which is the atmospheric layer found just beneath the stratosphere. An increase in blocking high patterns weather can inhibit the normal flow of the jet stream. When a weather system becomes blocked it remains stationary. These disruptive patterns can last up to two months and cold Arctic weather can be experienced at latitudes further south.

Conversely when the stratosphere is colder it leads to a stronger jet stream that locks the cold polar air mass safely over the Arctic. So a cold stratosphere can keep Great Britain bathed in a warmer jet stream air flow. While a SSW event makes Great Britain much colder. So paradoxically a sudden Stratospheric Warming Event causes a polar vortex air flow due to a kink in our jet stream.

The stratospheric warming leads to our westerly winds being suppressed and since the mild westerlies warm the British Isles this leads to a cooler air flow. When the wind comes from the east we have the "beast of the East" Siberian winds blowing. So there are two cold air patterns affecting the climate in Great Britain which are the north and the east winds. Conversely the west and southerly winds bring milder air flows.

Seemingly anomalous stratospheric warming effects indicate the complexities of atmospheric physics. A stratospheric warming can also cause the jet stream to reverse bringing cold eastern Siberian winds to the U.K. It is thought by scientists such as Mike Lockwood that such a cold easterly wind pattern may have prevailed during the last mini ice age.

March 1st 2018 saw the Siberian east wind bring biting cold and record amounts of snow. Indeed the cold east wind dominated the entire month and spring was late in arriving here in the British Isles. America suffered a similar fate and snow fell on April 1st in Washington DC, Philadelphia and Minnesota. These colder north east wind patterns will prevail when the ice age arrives in force.

Stratospheric warming events occur when large atmospheric waves, known as Rossby waves, extend beyond the troposphere where most of our usual weather occurs, and into the stratosphere. This vertical transport of energy can set a complex process into motion that leads to the breakdown of the high altitude cold low pressure area that typically spins above the North Pole during the winter. This is leads to the polar vortex flow. These undulating waves were first identified in the Earth's atmosphere in 1939 by Carl-Gustaf Arvid Rossby who went on to explain their motion.

When the Rossby waves hit the stratosphere the polar vortex is disrupted. What is a Rossby wave exactly? These waves are large ripples found in both oceans and atmosphere. A Rossby wave may consist of a high pressure ridge that is blocked from moving or disintegrating. They appear to be caused by the rotation of the Earth. This rotation is also known as the Coriolis force.

Rossby waves owe their origin to the tangential speed of the planetary rotation. This is a complex mathematical topic so I am keeping it very simple.Some scientists believe that the Rossby waves may be responsible for some of our extreme long lasting weather events recorded in the Northern Hemisphere. These stationary weather events have been characterised by a blocking high or anticyclones that have become stuck in one position for a long time. For example when there is a very long period of warmth as in the famous summer of 2003 this is due to a blocking high anticyclone weather pattern.

A similar stationary blocking high can cause a huge mass of cold air to stagnate for weeks bringing weeks of cold weather and ice. The weather front has in effect become blocked from moving and disintegrating so to speak and so remains stagnant. A blocking high pressure system or anticyclone can cause smogs if the air hangs over a polluted city for a long time.

A winter anticyclone can be clear and frosty or perhaps foggy and the air is still without any wind. A summer anticyclone can bring an endless succession of blue sky and sunshine filled days. Think of the unusually sunny and warm British summers of 1976 and 2003 to get an idea of how a blocked anticyclone weather pattern will cause a prolonged spell of weather.

Some meteorologists think that extreme weather events such as the widespread flooding event in Europe in 2013 and the 2012

China flooding are caused by a resonant amplification of Earth`s Rossby waves (Petoukhov, Ramstorf et al 2013). The following abstract of the Rossby wave paper states that,

"In recent years, the Northern Hemisphere has suffered several devastating regional summer weather extremes, such as the European heat wave in 2003, the Russian heat wave, the Indus river flood in Pakistan in 2010, and the United States heat wave in 2011. We propose a common mechanism for the generation of persistent longitudinal planetary-scale high-amplitude patterns of the atmospheric circulation in the Northern Hemisphere mid latitudes. Zonal wave numbers, m = 6, 7, or 8 are characteristic of the above weather extremes."

I have copied the transcript word for word so as to give an accurate representation of the theory. As one can see there is no mention of rising carbon dioxide levels to explain these extreme weather events. The events appear to be due to Earth`s rotational Coriolis effect and solar activity rather than to an anthropogenic warming caused by our carbon dioxide emissions. These days every single unusual weather event is blamed on carbon dioxide or greenhouse gases and this is rather naïve.

In the Northern Hemisphere the anticyclone rotates clockwise around the globe. In the Southern Hemisphere it rotates anti-clockwise. As mentioned previously these rotations are caused by the Coriolis effect of Earth`s rotation. These atmospheric waves are also responsible for dust storms in America. The British Met Office said that one type of jet stream factor known to affect weather is the quasi-biennial oscillation (QBO).

This cycle, discovered in 1959, involves a narrow band of fast moving winds like the jet stream which sits about 15 miles up over the equator. Waves known as gravity waves give rise to the Quasi Biennial Oscillation or QBO wind pattern. The QBO is believed to affect the ozone layer in the stratosphere. Eastward phases of the QBO often coincide with a sudden stratospheric warming and cold winters in northern Europe and eastern USA. Westward phases of the QBO coincide with mild winters in eastern USA.

The QBO correlates with a strong Atlantic jet stream. This brings mild and wet stormy winters to northern Europe and the British Isles. Such a wild wet weather system affected the British Isles in 2012.During the Queen`s diamond jubilee in the summer month of June, it rained so much that six people were taken to a London hospital with hypothermia!

The rains continued unabated for days and holiday campers had to be evacuated from flooded caravan parks in Wales. By September over 570 homes had been flooded according to the Environment Agency. The heavy rains and flooding continued until late November 2012 causing extensive damage and a lot of heartbreak as people surveyed their flooded homes. Some houses even collapsed and were washed into gushing rivers nearby.

During early spring 2012 the UK was positioned under a strong high pressure region resulting in very dry and warm conditions which brought a mini heat wave. Then everything changed in April, when the wave pattern underwent a significant shift. This shift put the British Isles at the mercy of strong low pressure weather systems. This low system brought relentless torrents of rain from the prevailing south-westerly flow and rivers burst their banks and bridges collapsed.

Motorists became trapped in their cars as the waters suddenly rose and a man died under a bridge. This relentless flooding was blamed on the jet stream by the Met Office. The wet weather was partly due to an aberration of the Rossby wave pattern according to meteorologists. The jet stream became disturbed in the Northern Hemisphere. This disturbance then resulted in a Rossby wave pattern of both high and low pressure regions.

The projected climate models for a global warming predict that our summers will become hotter and drier and grape vineyards would soon be flourishing here! However a lot of our weather here in Great

Britain is at the mercy of the prevailing pattern of jet streams. These in turn are influenced by planetary Rossby waves.

An eastward phase coincides with a weaker Atlantic jet stream and cold winters in Northern Europe and eastern USA whereas westward phases of the QBO often coincide with mild winters in eastern USA and a strong Atlantic jet stream with mild, wet stormy winters in northern Europe. In addition, the QBO has been shown to affect hurricane frequency during hurricane seasons in the Atlantic.

Research has also been conducted investigating a possible relationship between ENSO (El Nino Southern Oscillation) and the QBO. Will any loss of sea ice influence the Rossby wave pattern? Has the planetary Rossby wave activity changed in recent decades and is it likely to do so under a projected future warming? If it is changing, is the rapid Arctic warming indeed responsible? Is there a relation with climate change and Rossby wave patterns? These are some of the questions that are being asked by climate scientists James Screen and Ian Simmonds.

The Phenomenon known as Arctic amplification leads to a greater warming then anticipated at the Poles. As the Arctic warms it displaces cold air that then drops down over the British Isles. Arctic amplification refers to a far greater amount of warmth than should be expected from the actual amount of heat that has been convected from equatorial regions.

This phenomenon of the Arctic climate has been thoroughly researched by the distinguished and highly accomplished American oceanographer Dr. James Overland. It is akin to the famous butterfly effect chaos theory proposed by Edward Lorenz whereby a small change in an initial parameter can lead to a much large effect than anticipated.

Arctic amplification may also cause kinks in the path of the jet stream leading it to drop down instead of remaining as a tight circular belt around the North Pole. An increasing trend of cold Arctic air dropping over America and the British Isles is projected to increase in the future leading to bitterly cold winters. A team based at Sheffield University confirms that Arctic warming may affect the position of the jet stream resulting in cold winters across America and Northern Europe.

Many extreme weather events are actually correlated with jet stream activity and are not the result of carbon emissions. There is also a phenomenon known as Arctic amplification. This is the term given to the fact that a change in heat radiation will tend to result in an amplified change in temperature near the Arctic. So the warming at the Arctic or Antarctic will be far greater than that of the average temperature.

Rutgers University climate scientist Jennifer Francis and colleagues link a wavy jet stream to a warming Arctic. The team believes that climate changes near the top of the world are happening faster than in Earth's middle latitudes. Prof Jennifer Francis of Rutgers University in the US said that "This heavy rainfall does seem to suggest that weather patterns are changing and people are noticing that the weather in their area is not what it used to be".

Recently the atmospheric jet stream which transports weather systems has taken to meandering all over North America. This has resulted in prolonged cold snaps on the East Coast. Wavy jet stream patterns have been occurring frequently since the 1990s, and are now affecting weather around the Northern Hemisphere. Jennifer Francis is Professor at the Institute of Marine and Coastal Sciences at Rutgers University. She thinks that the new Arctic amplification warming pattern is causing extreme weather in America such as droughts and heavy snow fall.

Exceptionally cold winters are being reinforced by a meandering jet stream dragging down a cold polar vortex. This has caused some record snowfalls in America recently. The phenomenon known as Arctic amplification is caused by disproportionate Arctic warming that in turn is causing the Northern Hemisphere circulation to assume a more meridional wavy character. This phenomenon of

Arctic amplification might be causing highly amplified jet-stream patterns bringing freakish weather events (Francis & Skific 2015).

These strange events could actually be a sign of an ice age onset as there is evidence of similar Arctic warming before the end of the last interglacial known as the end Eemian. During an ice age the Arctic may paradoxically become warmer while the northern land masses become laden with snow and ice. There were large settlements of Inuit populations in the Arctic from the time of the last ice age, and the transit of people into North America could support this warmer Arctic hypothesis.

Most of Alaska was ice free during the last Ice Age and connected to Siberia by the now submerged land bridge called Beringia (Smith 2000). Early people migrated from Siberia to America across this land bridge. Interestingly to this day there is a strong facial resemblance between the ethnic communities of Alaska and Siberia.

Meridional jet streams can divert warm air into Arctic regions. In this way natural climate change gives us colder mid latitudes with more snow and ice cover during ice ages but strangely there may be warmer temperatures at the poles! It appears that the ice is somehow displaced and ends up covering Northern America and the United Kingdom. During interglacials we see warmer mid latitudes with less snow and ice cover and colder poles. Therefore the fact that the Arctic seems to be warming now could be a clear warning sign that a new ice age is arriving soon in the Northern hemisphere.

The reader may recall that the temperature on the graphs peak just before a rapid descent into an ice age. Recent warm decades provide the sort of data that one would expect to see just before the arrival of an ice age. Even the much publicised warming of the Arctic regions ties in with a predicted ice age event. In other words it always gets much warmer just before it gets cooler!

Conversely during interglacials we see warmer mid latitudes with less snow and ice cover and colder poles! So one can see that the scenario of an ice age is not entirely straight forward! Therefore the fact that the Arctic seems to be warming now could be a clear warning sign that a new ice age is arriving soon in the Northern Hemisphere!

Politicians of all political Parties have been milking anthropogenic climate change for all its worth. Climate change has provided an ideal excuse for their incompetence regarding the recent floods in Great Britain. During Storm Desmond over 16,000 homes were flooded. In December 2015 the Army evacuated 2000 residents from their homes and the Lake District was severely flooded in 2015. Many cases of homes flooding could have been prevented by proper Government investment in flood barriers and were probably avoidable. In 2014 the UK experienced a spell of extreme wet weather from late January to February as a succession of major storms brought widespread damage to the UK.

In December 2015 Leeds council leader Judith Blake accused the Government of discriminating between north and south when it comes to flood defence spending, saying that residents in her city were not getting "anywhere near the support that we saw going into Somerset" last year. Somerset used to flood regularly in centuries gone by.

Indeed the name Somerset derives from the word summer lands since Somerset was considered in days gone by to be only habitable during summer! This pattern of heavy rain fall was prevalent during the last Little Ice Age also. As the rains bore down on Somerset the Somerset levels became an island and residents were cut off by a moat! I suggested that our government make an application for assistance from the EU Disaster Fund.

The Government then proceeded to use climate change as a political comfort blanket to abrogate itself from any blame and incompetency. It was mischievously alleged that Lord Smith who was the Minister for The Environment had spent as much money on wining and dining in fine hotels as he had on dredging the rivers which had burst their banks! The departing environment Minister Lord Smith was feeling rather upset and so he decided to be cross with the hapless Lord Owen Paterson for not investing more money to adapt to climate change.

Accusing fingers were being pointed in all directions! An endless stream of Government Ministers such as Philip Hammond appeared on the news blaming the floods on climate change instead of a lack of Government investment in flood defences and infrastructure. It was all because of an act of God and that villain, carbon dioxide, that the rivers had burst their banks.

The blame was shrewdly shifted on invisible gases rather than the Ministers own incompetency. “Phew! This idea of climate change is proving remarkably useful chaps!” Now global warming could be held accountable for the endless mistakes and negligence of the ruling politicians. Now each and every flood caused by a lack of investment in maintaining rivers and flood barriers, could be blamed on the new fall guy, Mr. climate change!

The Met Office tried in vain to explain that the flooding had been due to the jet stream and that the jet stream is not affected in any way by greenhouse gases. However most of the time they simply went with the flow and tide of public opinion and said what the politicians wanted them to say. “Yes indeed we can expect to see more of this flooding as the climate changes in the coming future” the weather forecasters would say as they pointed to the flooded areas on the map of the British Isles. Of course this shifts the locus of responsibility away from the politicians who are running our country and gets them neatly off the hook.

David Cameron visited the flooded areas of York in 2015 where the River Ouse had burst its banks. He spoke with rescue workers and troops as he toured some of the worst-hit parts of the city while wading through water knee high in his green wellies. An angry reception of flooded homeowners heckled him and he replied that it was untrue that funding for flood defences had seen a 20% cut. The sodden Prime Minister promised that a major review of policy would look at whether more needed to be done and whether the strategy should be changed. York had been the focus of waterlogged controversy after the Environment Agency took the decision to open a flood defence barrier in an effort to prevent flooding elsewhere.

In fact the recent wet winters are remarkably similar to the wet winter endured by our British World War one war heroes in the trenches. A similar amount of rain fell on the troops during December 1914 to February 1915 filling the trenches with mud. The thought of our brave troops bogged down in muddy trenches saddens me. My own grandfather fought in the battle of the Somme. Suffice to say that the dreadful floods experienced by the British Isles were the result of a powerful jet stream that had brought a copious amount of water.

We have two wind patterns that are associated with bitterly cold weather in the U.K. and these are northerly and easterly winds. In contrast our southerly and westerly winds tend to bring warmer and milder weather. In England, prevailing west winds bring rain from the Atlantic. In winter, an east wind blowing from Northern Europe is always cold and often brings sleet and snow. Following a sudden stratospheric warming, the high altitude winds reverse to flow eastward instead of their usual westward. The eastward winds progress down through the atmosphere and weaken the jet stream, often giving easterly winds near the surface and resulting in dramatic reductions in temperature in Europe.

It is thought that during the Maunder Minimum brought about by low sunspot activity, there was a coincidental change in wind direction induced by the jet stream. So the British Isles had a continual cold air stream coming over from Russia in an easterly air flow which caused temperatures to plummet like a hail stone. So somehow there may be a connection between low sunspot activity and a change in our prevailing winter winds that serve to reinforce each other.

A recent paper was published in Nature Geoscience showing how low solar activity is correlated to jet stream blocking events. The authors looked at data going back for one thousand years for solar activity and climate. The study found a correlation between low sunspot solar activity and an increase in winter blocking anticyclone systems.

The team think that to the west of the British Isles there was a stubborn mass of air that blocked the warm westerly winds. This blocking high pressure system was being driven by the low solar activity. The researchers found that this anticyclone not only prevented the mild westerly air flow but it encouraged the colder air to dropdown from the North and East. This air pattern was behind the freezing winters of late in the British Isles (Nature Geoscience 2014).

This is the prevailing weather pattern that caused our country to become so very cold during the winters of the Maunder Minimum era of very low sunspots. Other meteorological studies have also confirmed that there is a relationship between solar variability and winter anticyclones and it is thought to be caused via the stratosphere. The lead author of the study in Cardiff University found a correlation with both the ocean and atmospheric dynamics during low sunspot activity.

The Welsh Cardiff University study was conducted in collaboration with the University of Bern. The team found that solar activity influenced the ocean-atmosphere dynamics in the North Atlantic Sea. This would lead to a vast effect on regional climate as the Gulf Stream would be affected. Professor Ian Hall said that they had used sediments taken from the seafloor in Iceland. The data was used to reconstruct both temperature and salinity of the seawater for the past 1000 years.

The study found that there were large temperature changes over time and that colder and less salty water was correlated with low solar activity. Professor Ian Hall was a co author of the study and he said that "By using the climate model it was also possible to explore how the changes in solar output affected the surface circulation of the Atlantic Ocean". "The circulation of the surface of the Atlantic Ocean is typically tightly linked to changes in the wind patterns". (Moffa-Sanchez et al 2014).

I find this research very interesting as it coincides neatly with recent findings by Bill Turrell on the Faroe Islands in Scotland about our Gulf Stream waters. Turrell noted that the salinity of the Gulf Stream is decreasing faster than at any time previously recorded. He is worried that it means our Gulf Stream is slowing down and may even stop altogether which would drastically affect our climate. I do not think that he has linked his findings with those of the solar physicists, and yet there is clearly a connection.

The academic focus has been directed on carbon dioxide emissions. Consequently this obvious connection between solar activity and our North Atlantic Ocean currents and Gulf Stream is being overlooked. It looks like the British Isles could soon be in a lot of trouble if the predicted low solar activity arrives in 2025. Solar physicists have also been recording a steady fall in the solar wind that will continue to decline until the sun goes into hibernation so to speak during the solar cycles 25 to 26.

This solar decline in turn could affect our Gulf Stream which is part of the North Atlantic Oscillation (NAO). A diminution of the solar wind could theoretically cause the Gulf Stream to slow down or stop completely. The North Atlantic Oscillation is a weather phenomenon in the North Atlantic involving atmospheric pressures at the sea level. The oscillation varies as a result of Icelandic lows and highs from the Azores down south. These differentials create oscillations and the effects are felt here in the British Isles as the westerly winds and currents can vary. The NAO is also part of the Arctic oscillation ocean current.

The winter of 2009–10 in Europe was unusually cold. It is hypothesized that this may be due to a combination of low solar activity and the jet streams (Science Daily 2010). The memorably snowy winter of 2010 was also observed to have a strong Easterly phase of the Quasi-Biennial Oscillation. This low solar activity resulted in the British Isles looking as if it had descended into an ice age. The satellite photograph below shows the exact amount of snow and ice cover that we would expect to see over the British Isles permanently during a new ice age.

The Met Office reported that the UK, had experienced its coldest winter for 30 years in 2009/2010 and again the coldest March day occurred in 2018. This cold also coincided with an exceptionally negative phase of the North Atlantic Oscillation. Thus we can see how low solar activity can affect the North Atlantic Oscillation and other ocean currents to bring very cold winter weather to the British Isles.

Some of our memorably wet winters were caused by jet stream anomalies and not carbon emissions please note! Jet streams are in turn caused by a combination of a planet's rotation on its axis and atmospheric heating by solar radiation. The jet stream is a band of fast moving westerly winds high up in the atmosphere which circle around the pole in the Northern Hemisphere. It can feature winds of up to 200 knots or more.

A strong westerly jet stream can bring wet and windy weather systems off the Atlantic and even cause flooding. The jet stream moves around and its position can have a big impact on weather here in the UK depending on where it is. If the jet is over the UK or just to the south, we tend to get a lot of wet and windy conditions as it brings weather systems straight to us. If the jet is to the north of us, it guides that changeable weather away to the north to leave the UK with more settled conditions. In centuries past cottages had reed matting on floors. When the floods came you moved all possessions upstairs and the matting could be discarded and replaced.

Great Britain may lose a tremendous amount of warmth when the winds change direction and the Gulf Stream slows. We will need more energy and this concerns me a great deal. If the solar physicists are correct then we should soon see a pattern of very cold icy winters starting to kick in. Biting cold north easterly wind patterns will prevail and we will need all the power and energy possible to keep ourselves warm.

As a small island we are at the mercy of prevailing wind and ocean currents. The Gulf Stream brings us free heat equivalent to a million power stations. If our sun goes to sleep, both the warm westerlies winds and the Gulf Stream will switch off. Therefore if solar activity switches both of these heat sources away from our vulnerable Isles, it is easy to comprehend how a little ice age will occur. The riddle of our climatic variables has been solved and it seems that we are in a precarious position.

The question dangling in the air like a shivering icicle is what will happen to the benign climate of the British Isles? There is no certainty but a sense of foreboding looms. We are in a vulnerable position. The problem is that Great Britain lies much farther north than most people realise. The odds are all tipping us towards another ice age. While we feel safely cuckooned in our comfy bubble that is heated by the Gulf Stream and mild westerlies, an ice age looms. A prolonged kink in our jet stream may bring us cold east winds or an Arctic blast that one day may precipitate our ice age. If I was a gambler my bet would be that the future of the British Isles is not one of fragrant vineyards but one that is covered in ice!

Chapter 13: Killer Ice Storms and Snowblitz

Ice storms cause a build up of ice on power lines and trees leading to a collapse of these structures in a short time. America and Canada have been hit with many ice storms in recent years. Canada was hit by one of the worst ice storms of the century in 1998 on January 5th. The ice storm covering Quebec and Ontario was a national disaster. When the Canadians awoke they heard the cracking sounds of branches snapping. Tens of thousands of trees and branches snapped in rapid succession damaging power lines and blocking roads.

Branches laden with ice were falling out of the sky every second. Then the blackouts came and the power went down. Then the giant electricity pylons began to collapse under the weight of ice. Around 600, 000 people had to be evacuated from their homes into shelters. A total of 955 people were injured by the ice storm and 35 fatalities. At least 300, 000 animals perished from cold. The maple crops were destroyed. The water filtration plants lost power leading to water shortages.

The army was deployed to clear the trees and many people lost their lives in the killer ice storm. As well as the human life lost there was a massive bill for insurance companies totalling billions of dollars and lost productivity as two million employees were unable to get to work. There was a domino effect as the wooden utility poles collapsed so that one by one the power network collapsed like a deck of cards.

Metal pylons bent over at strange angles and the whole scene resembled a disaster movie. The storm caused complete and utter chaos. Five million residents were left without power for days and had to dine in cold food by candlelight. Such ice storms are far from rare in Canada and America and these life threatening storms will increase as the ice age bites. The 1998 ice storm of Quebec is a taste of things to come.

Ice ages are still shrouded in mystery. The truth is that we really do not know exactly how a major ice age arrives. Perhaps the ice slowly accumulates in stages until a mile thick ice sheet covers most of the British Isles. Or perhaps the ice age arrives with a bang not a whimper and ice storms arrive with a vengeance! A giant prehistoric hail storm is thought to have given rise to a prehistoric mini ice age.

According to Professor Napier from Cardiff Astrobiology Centre, the Earth strayed into the path of a shattered comet and this caused a massive hail storm. This event possibly triggered the Younger Dryas ice age 13, 000 years ago. Certainly the freeze dried mammoth remains are compelling evidence of a sudden ice age onset. Some climatologists speculate that sudden and heavy accumulations of snow may arrive in the Northern Hemisphere in the space of years rather than centuries. The cooler temperatures from the lazy sun may then prevent the snow from melting.

An ice age covering the Northern Hemisphere is a highly probable event. There is already sufficient snow and ice in the Arctic to precipitate an ice age in Great Britain. The air masses above the North Pole are also sufficiently frigid and at previous ice age temperatures. All that is needed is a prolonged displacement of this frigid air further south as has happened recently in the polar vortex winters. When the cold Arctic air is dragged south by the jet stream we see large accumulations of snow and ice. The jet stream is in turn affected by sunspot solar variables.

Other apocalyptic scenarios may involve a huge precipitation of snow and ice that lasts for weeks on end until houses are completely buried. What might cause such a snowblitz event? One possibility might be that the underwater Arctic volcanoes become active releasing warming lava and gases into the frigid seas. This theory is from the book "Not by fire but by Ice "by Robert Felix. This is not as farfetched as it sounds when one considers the hot volcanic springs in Iceland.

The cold seas would warm up to sufficiently facilitate evaporation into the very cold skies above. When the warm moisture meets the frigid air a heavy precipitation of snow could result. The polar vortex could then drag this heavy snow over the British Isles and bombard us with weeks of snow. A more prosaic possibility is the "Beast from the East" cold Siberian wind, may lock over Great Britain bringing cold biting winds. When this cold air mass collides with a stormy wet weather system it could bring a huge precipitation of snow.

As our magnetic field continues to weaken we will be bombarded with both solar coronal mass ejections and cosmic rays. The CME`s would arrive in force as the magnetic field declines. Undeflected solar CME events could theoretically cause a huge evaporation over the Arctic Ocean. As the evaporated sea water meets the cold air masses above it could lead to a snowblitz event. The cosmic rays could also facilitate a snowblitz by increasing clouds that will fill with snow. Remember that cosmic rays seed clouds and are proven to decrease global temperature. This deadly space weather would wreak havoc on Earth as our magnetic shield goes down. As the magnetic pole shifts the seismic activity would go through the roof. Many volcanoes will erupt cooling the planet.

These are some of the probable scenarios that will help initiate the ice age snowblitz. Such a snowblitz would result in an ice age arriving within months. It seems highly probable that the weakening of our magnetic field may be the pivotal factor along with a weakening sunspot cycle. As the sun becomes quiescent the temperatures will fall further. When the colder air meets the moisture there will be a snowblitz. Already there have been heavy snowstorms all over the Northern Hemisphere and even as far south as Greece and Arab states.

When it snows and hails for days on end the infrastructure of unprepared cities may soon grind to a halt. When an ice storm hit Canada in 1998 over 1000 transmission towers collapsed. Ice rained down for 80 hours without a break! There was chaos as electricity pylons collapsed and folded under the weight of the ice. Four million Canadians were left without power and there were many deaths from hypothermia.

The water stations were also disabled in Montreal and bridges were closed. Pedestrians were endangered as large chunks of ice fell from the rooftops Large areas of Old Montreal were cordoned off by police to protect pedestrians from blocks of falling ice . Millions of trees were brought down by the weight of the ice. Quebec`s world famous maple sugar crops were devastated and the apple orchards were ruined. Fourteen thousand troops were deployed to help with security and evacuation. Snowstorms have also battered New York many times.

Snow storms are now increasing all over the world as we gear up for the ice age. In September 2008, Tibet recorded its worst ever snowstorm. As the snow fell thousands of people became trapped in Lhunze County. Many buildings collapsed and cattle were buried alive. China was also affected by a wave of killer snowstorms in 2008 resulting in many fatalities. It has even snowed recently in the Taklamakan Desert in western China.

Hypothermia, frostbite and wind chill exposure can cause people to become disorientated and confused. The elderly are especially at risk. Pensioners are less inclined to exercise and may live a more sedentary lifestyle perhaps due to arthritis or back problems. Blizzards are especially lethal as the snow is combined with very high strong winds. Hypothermia is a medical emergency and happens when the body temperature falls below 35 degrees Celsius.

There are various stages of hypothermia starting with shivering as the body attempts to warm up. As the hypothermia advances to a more severe state, shivering may stop. Age U.K. has reported that more than five million households are suffering fuel poverty and many elderly people face a harsh choice of eating versus heating. There are an excess of around 40,000 extra winter deaths reported every year due to the cold.

This is a sad irony as heating bills have risen to pay for global warming measures. Grants are supposed to be available to help insulate cold homes and yet many homes are slipping through the safety net and pensioners are still dying from hypothermia and cold related illnesses. Whilst there are thousands of extra deaths related to the cold every winter, how many comparable deaths do we see here in Great Britain from global warming? Common sense should indicate where our focus should be directed.

The recent snowy winters in the British Isles caught us by surprise. Britain usually benefits from mild winters thanks to warm westerly winds and the heat gifted by the Gulf Stream. The same cold weather also hit America, Europe and China. This biting cold was blamed on the polar vortex. More recently in early January 2017 and 2018 the same old villain the polar vortex reappeared bringing freezing cold and snow to many parts of Great Britain and America.

As discussed previously the polar vortex is a mass of freezing air that usually sits above the Polar region. We are mainly concerned here with the Northern polar vortex as this is the one which brings havoc to our shores. The polar vortex rotates in an anticlockwise direction at the North Pole. The polar vortex varies in strength from time to time. When it starts weaken it somehow diverges into two air streams or vortices. The dangerously cold polar air mass is then brought down to cover Canada, America and Europe.

The weather forecasts by our Met Office frequently mention how the polar vortex is able to affect the position of our jet stream. The polar vortex is able to push the jet stream into a north/south kink, thereby causing the freezing cold polar air to descend into this kink or dip. A record breaking polar vortex hit America in 2014. It was so cold that even a zoo`s polar bears could not cope with the cold! Polar bears at a zoo in Chicago needed to be brought indoors. It seems that the polar vortex winter phenomenon is becoming more frequent as we are seeing an increase in snowy winters.

In February 2017 a news report announced that there is a shortage of lettuce here on supermarket shelves. The reason for this shortage is that Spain had unusually cold weather. There have been heavy rains and snow in countries that are normally warm and sunny. Many vegetable crops in southern parts of Europe have been buried under unexpected snowfalls and therefore ruined. There have been heavy snows all over Southern Europe including Greece and Turkey. There are reports of excited children who have never seen snow before in their young lives playing with snowballs!

Many British pensioners like to escape to the warm Southern countries of Europe during the British winter. Well the recent cold snowy weather over Southern Europe might prove a disappointment to the expat escapees, if it continues to follow this pattern! We are regularly seeing reports of vast swathes of Europe buried in snow in countries that normally never see a snowflake.

If I were a gambler my money would be placed on the arrival a major ice age event soon rather than an extended period of balmy warmth. The solar sunspot minimum is not expected yet for a decade and we are already witnessing worrying signs of heavy snowfall and unusually cold weather over Europe. Could this be the onset of a new major ice age rather than the expected mini ice age?

I vividly recall chatting to an Oxford friend called George Marshall in Oxford back in 2007. George Marshall is well known for his work leading the environmental charity COIN, which is the Climate Outreach Information Network based in Oxford. George Marshall is a very likeable chap and gives very entertaining talks in America and England on how to become a climate champion. The COIN charity is thriving and makes people aware of how to curb their carbon footprints.

I remember that we were having yet another mild winter and George told me earnestly that we may never see snow again in the U.K. I remember feeling rather disappointed at the forecast since I rather like the look of snow and enjoy making snowmen, having snowball fights and all the other fun snow activities! Then as luck would have it there followed a succession of snow laden winters! However this is not a laughing matter as cold weather may be life threatening to the elderly. Still it showed that predictions of a warmer and grape growing climate here may be incorrect.

Another climate scientist at the Climate Research Unit, CRU, named David Viner said in the year 2000 that children today may never live to see snow again during their lifetimes. Then as fate would have it, the heavy snows started to return once more to our temperate island. Suddenly snowy winters were back in vogue and it seemed that the projected snow free winters had perhaps been a miscalculation. The expression famous last words, springs to mind! There has been plenty of snowfall in the British Isles since the snow free weather projection was made!

The increasing snowfall that we are seeing here in Great Britain could be a warning sign that we are ill advised to ignore. Of course a few snowy winters do not mean by themselves that an ice age is coming. Our winter weather here is notoriously unpredictable due to our four air masses that govern our weather. When our mild westerly winds prevail the winters are mild and wet and even stormy. This weather pattern usually brings a mild but wet winter. If however this mild air flow were to meet a cold northerly air flow then we might have a lot of snow.

The change in wind patterns can make a huge difference to the kind of winter we may experience here in the British Isles. Our dominant wind pattern is for the mild south westerly winds to prevail. This is the main reason for our relatively mild winters. If this prevailing wind pattern were to change, our winters would be a lot colder. As the old proverb goes "a north wind doth blow bringing sleet and snow". The British Isles has a climate that is a lot milder than it should be for its latitude, thanks to mild maritime air masses from the west and south.

In February 2018 the Met Office announced that a prolonged spell of freezing weather over the British Isles was due to a strange phenomenon known as a sudden stratospheric warming. This phenomenon of a sudden stratospheric warming over the North Pole paradoxically brings freezing cold air downwards. During a SSW event the air in the stratosphere collapses and warms. The stratosphere is ten miles above the surface of the Earth.

When the SSW event occurs it leads to wild swings in the jet stream and air pressure. The normally tight band of jet stream around the North Pole becomes kinked and drags cold air from the north and the east. Cold east winds from Scandinavia are directed to the British Isles as the jet stream deviates from its polar circular flow. The main effect of this sudden stratospheric warming is that the mild westerlies are stopped and are replaced by cold wind patterns.

A blocking high weather system of cold stationary weather then forms. This anticyclone locks the cold air tightly in place. So one can see that there is often a displacement effect of weather, whereby the cold Arctic air is dragged south to cover the British Isles. In late February 2018 the cold east wind known as the "Beast from the East" was dragged over Great Britain. Then the cold Siberian air mass was joined by another air mass from the south west bringing lots of moisture. The result of the cold and wet air masses colliding was a whiteout of snow over the entire country.

Despite warnings Great Britain was completely unprepared for the coldest March day since records began. On March 1 2018 cars became buried in deep snow drifts. Drivers who had ignored warnings and taken to the roads became stranded overnight in freezing temperatures. Even the gritters could not get through. Trains ground to a halt and commuters walked on the rails to escape! The problem arose because of the overloaded capacity on our stretched transport system. Our roads are already running at a peak capacity. Therefore it only takes a small hiccup for the entire transport system to collapse and become gridlocked.

This snowpocalypse event in 2018 cost the British economy £1 billion per day due to missed work and insurance claims. There were 8000 reported collisions on the roads. Schools were closed and hospital operations were cancelled. It was reported on the news that a doctor had walked ten miles to get to his patients, while other hospital staff slept in the hospital. Transport ministers then nonchalantly announced that these rare snowblitz events did not make it worthwhile investing in more snow ploughs!

These snowy weather events make it possible to imagine how the ice age might start in the future. In Great Britain we are at the mercy of four wind patterns and they could result in a huge precipitation of snow in a short while. An ice age might arrive extremely rapidly and dramatically in a wave of furious storms called a snowblitz. These storms produce a rapid accumulation of snow that starts to accumulate without melting. This in turn produces a greater albedo that results in a further cooling .Instead of taking centuries to arrive, the ice age could hypothetically arrive in months.

One should remember that evidence of a very sudden onset of glacial conditions has been found in pollen records. Plant expert Genevieve Woillard found changes in boreal zones that demonstrated a very rapid onset of ice age conditions, taking place over decades rather than centuries. Therefore it seems unlikely that an ice age will creep upon us slowly from the Arctic Circle over a thousand years. Interestingly during the last ice age there was a similar amount of ice at the North Pole as there is at the present day. This finding shows that we really are not too far off ice age conditions. All that is needed is a succession of heavy snowfall events caused by colliding air masses.

Further proof that ice ages begin abruptly has come from the Greenland Ice Core Project (GRIP). Drilling cores two miles deep, GRIP has obtained a climate record for the past 250,000 years. It seems that the last ice age began catastrophically. Worldwide the temperature plummeted by 20 degrees Fahrenheit almost overnight (Anklin et al 1993).

Violent isotopic changes have been recorded. The sudden isotopic changes prove that an instantaneous ice age took place around 90, 000 years ago. According to strontium ratios it appears that every ice age began promptly and dramatically! So when the ice age arrives it will come like a speeding express train! Hopefully it will not arrive overnight or we may wake up one day to find our homes are buried in sonw! This isotopic record gives rise to the possibilty of a mega snow storm where it snows several feet a week until everything is buried beneath a ton of snow twenty feet high! It is just possible that this is how our next ice age will arrive.

The snowblitz theory was first popularised on a programme called the Weather Machine in 1975. At that time there was more concern that a new ice age might arrive than about global warming. This might be because there was a general cooling of weather in the U.K. between 1940 and 1970. Might a snowblitz lead to a sudden ice age?

Picture Credit; Sun Gazing February 2013

The above picture shows hundreds of vehicles buried in snow while driving along a highway in Suffolk County, Long Island, NY. This is Route 25 and shows the highway looking like a car park rather than a busy route. Stranded motorists were rescued by snowmobile after nearly 30 inches of snow simultaneously fell in one giant amount! It must have been quite an adventure for the motorists! Fortunately, there are now superb engineering advances being made in snow clearing machines. We would be wise to invest in these new machines and snow clearing companies.

We are already seeing increasing snowfall all over the Northern Hemisphere and as far south as Spain, Greece, Turkey and Italy. America has been hit by several massive snowstorms such as storm Stella in March 2017 which caused major disruption and the cancellation of thousands of flights. In 2018 many snowfall records were broken as the USA suffered a dramatic snowblitz. In Boston temperatures were recorded at minus 19 C.

Even Southern Europe was plunged into a polar vortex in 2017 and snow appeared on Italian and Greek beaches! The Adriatic Sea in Italy recorded temperatures of minus 20 degrees Celsius. The increased snowfall events are warning signs that an ice age looms ever closer and we ignore the omens at our peril. A snowblitz of hail and snow could one day rain down on us here in Great Britain for weeks on end and lead to a new ice age.

In January 2018 temperatures were recorded as 30 degrees below average in North Eastern America and snow was piled up to 90 inches deep in places. More than a million Americans homes were left without power for days. In March 2018 snow fell in Morocco and North Africa at a time when the temperature should have been climbing into the low seventies! These snow storms are just a small taste of things to come. When the expected ice age arrives the temperature could be up to 40 times colder here in Great Britain and America. Welcome to ice age Britain!

Conclusion: Forewarned is Forearmed.

It was reported that some wealthy Californian businessmen have bought acres of land in New Zealand to prepare for the coming ice age. They have stockpiled huge amounts of tinned food in caves and built themselves comfortable and sophisticated subterranean dwellings! Clearly they are expecting an apocalyptic event and envision food shortages and anarchy. Are they over reacting? Certainly they are wise to go off grid and become self sufficient.

There is a thriving prepper movement in America consisting of at least three million worried people who are preparing for some sort of apocalypse. Americas is a good place to be in an apocalypse as it has plenty of land and woods where people can become self sufficient. Preppers also tend to stockpile ammunition that they see as essential for hunting animals for food. Many ordinary Americans such as financiers are now secretly building up an arsenal of weapons in their basements.

These survivalists are making preparations to survive the collapse of a society based on a fragile technology and they have stockpiled tins of food. Some have fish in their ponds to keep them nourished if the food supply collapses. This collapse could happen if a massive solar flare Carrington event arrived knocking out the power grids. Such CME events may occur as our ozone layer and magnetosphere continue to weaken.

The more technologically advanced a civilisation, the greater a potential descent into anarchy. Any society using technology that depends on binary code is at risk of collapse into complete anarchy. If a major ice age arrives, then those who live a basic self sufficient lifestyle will be in the best position to survive. If an ice age arrives suddenly, then those who live with their own power supply and source of food will also fare better. If a major ice age arrives, then those who live a basic self sufficient lifestyle will be in the best position to survive.

Great Britain is in a very precarious position as we are so near to the Arctic Circle. It seems that all of our Northern Hemisphere climatic feedback loops are conspiring towards an ice age for Great Britain! There is very little likelihood therefore of a warm balmy Mediterranean climate returning to Great Britain. This is a prediction that climate modellers have got completely wrong. The climate feedback loops will always result in Great Britain *becoming cooler rather than warmer*. This is why the title of my book is "Ice Age Britain".

The general public has been led to believe that the main drivers of our climate are the greenhouse gases. In fact the main climatic drivers are as follows: solar activity, the Coriolis force, jet streams, the hydrosphere, wind patterns, ocean circulation patterns, Milankovitch orbital variables, cosmic dust, sulphur aerosols, volcanic eruptions, cosmic rays and cloud cover, to name a few! Phew!

On a serious note, I wrote this book not because I hanker for fame but as a matter of national security. We must start to prepare for the ice age that will inevitably arrive on our shores. In the United States there are up to three million survivalists who are prepared for a collapse of our modern day society. When the national grid collapses, they will burn the wood from their land to cook and keep warm. When the infrastructure to deliver food breaks down, the prepared survivalists will have home grown food and will hunt for wild animals.

The Earth has been steadily cooling down over the last 60 million years and the last two million years have been the coldest ever recorded! Since Earth is on a multi-million year downward cooling trend, we may get the mother of all ice ages. The next ice age may be even deadlier than the last ice age when the world population shrunk to only a few million. We will soon be reaching the end of the safe Holocene. Once more we will plunge into an ice age so severe that it may even spell the end of humanity. You have been warned!

Acknowledgements:

Abe-Ouchi A., Saito, F., Kawamura K., Raymo M., Okuno J., Takahashi K., Blatter H., Insolation-driven 100,000-year glacial cycles and hysteresis of ice-sheet volume *Nature* 500 p 150-193 8 August 2013

Ackerman, A. S., et al. (2000). "Effects of Aerosols on Cloud Albedo: Evaluation of Twomey's Parameters of Cloud Susceptibility Using Measurements of Ship Tracks." *Journal Atmospheric Sciences* **57**: 2684-95.

Alvarez et al, (1980), "Extraterrestrial cause for the Cretaceous –Tertiary extinction", *Science,* 208 (4448):1095-1108)

Asch, S.E. (1951). "Effects of group pressure on the modification and distortion of judgments" in H. Guetzkow (Ed.) *Groups, leadership and men* (pp.177–190) Pittsburgh, PA: Carnegie Press.

Alvarez, W. et al (1980)"Extraterrestrial cause for the Cretacious-Tertiary Extinction" *Science* vol. 208 pp 1095-1108.

Anklin, M. et al "Climate instability during the last interglacial period recoded in the GRIP ice core *Nature* vol 364 pp203-207 15 July 1993(GRIP) 1993

Arrhenius, Svante "On the Influence of Carbonic Acid in the Air upon the Temperature of the Ground" *Philosophical Magazine and Journal of Science* Series 5, Volume 41, April 1896, pages 237-276.

Behringer, W. (1999): Climatic Change and Witch-hunting: the Impact of the Little Ice Age on Mentalities. *Climatic Change*, Vol.1(1): 335-351

Bell, A. and Strieber, W. (2000).*The coming global superstorm*, Pocket Books, New York,

Betts, A. K. and Ball, J. H., "Albedo over the boreal forest" *J. Geophys. Res.* **102, D24**, 28901–28909 (1997)

Biello, D. "Did a Comet hit Earth 12000 Years Ago". *Scientific American* Jan 2 2009.

Biello, D. (July 18, 2012). "Controversial Spewed Iron Experiment Succeeds as Carbon Sink". *Scientific American* July 19, 2012).

Bond, G., et al (2001) "Persistent solar influence on North Atlantic climate during the Holocene". *Science* 294: 2130-2136

Bowen, D.Q., (1978) *Quaternary geology: A stratigraphic framework for multidisciplinary work*, Pergamon Press Oxford, New York,

Box, S. (1983) *Power, Crime and Mystification*, London: Tavistock.

Broecker, W.S. "Unpleasant surprises in the greenhouse?" *Nature*, vol. 328, pp. 123-126, 1987

Bryden, H. L. et al "Atlantic currents show sign of weakening" *Nature* **438**, 655-657 (December 1 2005) .

Calder, N. "The Arithmetic of Ice Ages", *Nature,* Vol. 252, pp. 216-18, 1974

Calder, N. 1999 **scientific paper: *'The Carbon Dioxide Thermometer',*** *Energy & Environment,* 1999, Vol. 10, pp. 1-18

Cardiff University (2016, October 26). "Why does our planet experience an ice age every 100,000 years?" *Science Daily* Retrieved February 22, 2017

Carlson, A. E. (2013) "The Younger Dryas Climate Event" *Encyclopedia of Quaternary Science* **3** Elsevier pp.126–34.

Carrington,D. "Britain`s damp, leaky homes among most costly to heat" *The Guardian Newspaper* 29 November 2013

Casey, J.L. "Cold Sun" (2011) Trafford publishing

Castro, J. "Mountains may suck up carbon better than thought", Jan 16 20 *Live Science Online*

Champion, D. E., Lanphere, M. A., and Kuntz, M. A., 1988, "Evidence for a new geomagnetic reversal from Lava flows in Idaho: Discussion of short polarity reversals in the Brunhes and late Matuyama polarity chrons," *Journal of Geophysics Res.*, Vol. 93, p. 11667-11680, 10 Oct. 1988

Charlesworth, J. K. (1957) *The Quaternary era*, Edward Arnold, London

Clark, W. 1974, "Energy for Survival, the Alternative to Extinction", Better World Books

CLIMAP (1981) "Seasonal reconstructions of the Earth's surface at the last glacial maximum" in *Map Series*, Technical Report MC-36. Boulder, Colorado: Geological Society of America.

Cohen, J., et al "Recent Arctic amplification and extreme mid-latitude weather" *Nature geoscience* 7, 2014 pp 627-637

Crowther,T.W., Stephen M Thomas, Daniel S Maynard, Petr Baldrian, Kristofer Covey, Serita D Frey, Linda TA van Diepen, Mark A Bradford "Biotic interactions mediate soil microbial feedbacks to climate change" (2015/6/2) Proceedings of the National Academy of Sciences PNAS vol 112, issue 22 pp 7033-7038

Dansgaard, W. et al. (1993) "Evidence for general instability of past climate from a 250-kyr ice-core record". *Nature* **364** (6434): 218–220

Davies, R. (Oct 7 2016) *the Guardian* **"**Kite power to take flight in Scotland next year"

Day, S. J; Carracedo, J. C; Guillou, H. & Gravestock, P. (1999) "Recent structural evolution of the Cumbre Vieja volcano, La Palma, Canary Islands: volcanic rift zone re-configuration as a precursor to flank instability"*J. Volcanol. Geotherm Res.* 94, 135–167.

Dowdeswell and White, "Greenland ice Core Records and Rapid Climate Change" *Philosophical Transactions: Physical Sciences and Engineering* Vol. 352, No. 1699, *The Arctic and Environmental Change (*Aug. 15, 1995), pp. 359-371

Dowdy, Andrew; et al (2007) "Polar mesosphere and lower thermosphere dynamics: Response to sudden stratospheric warmings". *J. Geophys. Res.* **112**:

Elliot, M., Labeyrie, L., and Duplessy, J.C. 2002 "Changes in North Atlantic deep-water formation associated with the Dansgaard-Oeschger temperature oscillations (60-10 ka)". Quaternary Science Reviews 21: 1153-1165.

Evan, A. T., and Norris, J. R., (2012): On global changes in effective cloud height. Geophys. Res. December 2003 "Observed global cloud and radiation flux changes since 1952", American Geophysical Union

Eddy, J. A. (June 1976) "The Maunder Minimum" *Science* 192.

Kelso, J. and Danneman, M., 2017 "The contribution of Neanderthals to phenotypic variation on modern humans" *American Journal of Human Genetics* vol, 101: pp, 1-12

Kerry E.; et al (July 20, 1995) "Hypercanes: A Possible Link to Global Extinction Scenarios". *Journal of Geophysical Research* **100** (D7): 13755–13765.

Erickson, J. (2001) "Lost Creatures of the Earth: Mass Extinction in the History of life" Facts on File publishing books, New York

Europa (April 6 2006) "EU climate change policies: Commission asks member states to fulfil their obligations".

Felix R., (2000) "Not by Fire but by Ice" Sugarhouse Publishing

Fleming J.R. (2005) "Historical Perspectives on Climate Change". Oxford University Press pp. 69–70(Details of Tyndall's device for measuring the infrared absorptive power of a gas)

Fitzgerald, R.G. & Gleadow, A.J.W., 1988, Fission-track geochronology, tectonics and structure of the Transantarctic Mountains in northern Victoria Land, Antarctica. *Chemical Geology* (Isotope Geoscience Sections), **73,** 169-98

Francis, J.A., Vavrus, S.J. "Evidence for a wavier jet stream in response to rapid Arctic warming." 2015 *Environment Research Lett.* 10 014005

Francis, J. and Skific, N. “Evidence linking rapid Arctic warming to mid-latitude weather pattern *Philo Trans A Math Phys Eng Sci.* 2015 Jul 13; 373(2045): 20140170.

Frolich C., “Solar Irradiance Variability since 1978. Revision of the PMOD Composite during Solar Cycle 21”*Space Science Reviews,* Volume 125, Issue 1-4, pp. 53-65, August 2006

Ganopolski A. et al; *Nature* **529**, 200–203 (14 January 2016)

Glatzmaier, G.A. and Roberts P. H 1995 “a three dimensional self consistent model of a geomagnetic field reversal” *Nature.* 377 pp 203-209

Gosden, E. Jan 16 2015 “Labour energy price freeze 'preventing £130 bill cuts; Consumers are stuck paying high gas and electricity prices as a result of Ed Miliband’s policy, ” *the Telegraph*

Greenpeace Press Release March 4, 2016 “Fukushima nuclear disaster will impact forests, rivers and estuaries for hundreds of years”.

Gribbin, J. (1990)”Hothouse Earth; the Greenhouse effect and Gaia” Bantam press

Gribbin, J. (1988)”The Hole in the Sky; Man`s Threat to the Ozone Layer”, a Corgi book by Transworld Publishers Ltd

Hand,E. “Crippled Atlantic currents triggered ice age climate change” *Science* June 30 2016

Hansen,J. (2010) “Storms of my grandchildren” publisher Bloomsbury, N.Y.

Haug, G.H., “How the Isthmus of Panama put ice in the Arctic” *Oceanus Magazine* 22 March 2004, 2017

Hays J.D. (1976) “Variations in the earth's orbit: Pacemaker of the ice ages,” Science, vol. 194, pp. 1121–1132.
Hays, J.D. Imbrie, J.; Shackleton, N.J. (1976) "Variations in the Earth's Orbit: Pacemaker of the Ice Ages". Science ***194*** *(4270): 1121–1132.*

, **Henry L. G., McManus J. F., Curry W. B., Roberts N. L., Piotrowski A. M., Keigwin L. D**., North Atlantic Ocean circulation and abrupt climate change during the last glaciation, *Science* 29 July 2016: Vol. 353, Issue 6298, pp. 470-474

Higgins et al, *“Sediment focusing creates 100-ka cycles in interplanetary dust accumulation on the Ontong Java Plateau,” Earth and Planetary Science Letters 203 (2002) 383-397*

Hoyle, F. “ice the ultimate human catastrophe* Amazon 1979, 1983

Howard, B. C., *“Bizarre earthquake lights finally explained” National Geographic Jan 7 2014*

Imbrie, John and Imbrie, Katherine Palmer, (1979) Ice Ages: Solving the Mystery Enslow Publishers, Short Hills, NJ

Imbrie,J.et al. *“On the structure and origin of major glaciation cycles, linear responses to Milankovich forcing” Paleoceanography* **7**, 701–738 (1992).

Jackson, L. C., Peterson, K. A., Roberts, C. D. and Wood, R. A.; “Recent slowing of Atlantic overturning circulation as a recovery from earlier strengthening”; *Nature Geoscience* (2016).

Jaworowski, Z., 1994, *Ancient atmosphere - validity of ice records,* Environmental Science and Pollution Research, **1**(3): p. 161-171.

Jaworowski, Z., T.V. Segalstad, and N. Ono, (1992) "Do glaciers tell a true atmospheric CO_2 story" *The Science of the Total Environment*, **114**, p. 227-284.

Kennet, J.P. and Watkins, N. D., "Geomagnetic polarity change, volcanic maxima and faunal extinction in the south Pacific" *Nature*, Vol 227, p 930 -9934 , August 29 1970.

Kenyon, Scott J., Benjamin C. Bromley (2004) "*Stellar encounters as the origin of distant Solar System objects in highly eccentric orbits*" *Nature* **432** (7017): 598–602.

Khaitovitch et al (*Nature* online communications, April 2014).

Kerry et al "Hypercanes: a possible link in global extinction scenarios, *Journal of Geophysical Research*, vol.100,pp 13755-13765, July 20 1995

Kirby K. J. et al (2016). "Long-term changes in the tree and shrub layers of a British nature reserve and their relevance for woodland conservation management" *Journal for Nature Conservation*, 31, 51-60

Komitov, B., and Kaftan Vladimir "The sunspot cycle no. 24 in relation to long term solar activity" *Journal of Advanced Research* March 30 2013

Kopp, R.E. "The Paleoproterozoic snowball Earth: A climate disaster triggered by the evolution of oxygenic photosynthesis", Division of Geological and Planetary Sciences, California Institute of Technology 170-25, Pasadena, CA 91125; Communicated by Paul F. Hoffman, Harvard University, Cambridge, MA, June 14, 2005

Kukla, G.J. Matthews, R. J. (1972) "When will the present interglacial end?" *Science* **178 (**4057) 190-202

Kukla,George J., et al (1972) "The end of the present Interglacial." *Quaternary Research* **2**: 261-69.

Lamb, H. H. (1995) "Climate history and the Modern World" Routledge publishing

Lee, X. et al "Observed Increase in Local Cooling Effect of Deforestation at Higher Latitudes," *Nature* 17 nov. (2011). *Nature 479 pp 384-387*

Leyser T. B. and A. Y. Wong "Powerful electromagnetic waves for active environmental research in geospace" (*Reviews of Geophysics*, Vol. 47, RG1001, 2009).

Lovelock, J., Epton, S, "The Quest for Gaia", *New Scientist*, 65, 934:304. Feb 6 1975

MacDonald, Francis, *Calibrating the Cryogenian,* Science, 5 March 2010: Vol. 327 no. 5970 pp. 1241-1243 5 March 2010

Malley, C.S., Johan C.I. Kuylenstierna, Harry W. Vallack, Daven K. Henze, Hannah Blencowe, Mike R. Ashmore. "Preterm birth associated with maternal fine particulate matter exposure: A global, regional and national assessment". *Environment International,* Feb 10 2017

Mann, M.E. (April 1 2014) "Earth will cross the climate danger threshold by 2036", *Scientific American.*

Markham,D. "5 hour energy creator to roll out pedal powered energy solution in India" *Treehugger blog* Oct 6 2015

Marvel, K., G.A. Schmidt, R.L. Miller, and L. Nazarenko, 2016: Implications for climate sensitivity from the response to individual forcings *Nature Climate Change*, **6**, no. 4, 386-389,

Mc Guire, B. (1999) Apocalypse, a History of Natural Disasters World books ltd.

Melamed M L; Michos E.D.; Post W.; Astor, B. (2008) "25-hydroxyl Vitamin D Levels and the Risk of Mortality in the General Population" *Arch. Intern. Med.* **168** (15): 1629–37.

Mesnage R., Renney G., Seralini G. E., Ward M., Antoniou M."Multiomics reveal non-alcoholic fatty liver disease in rats following chronic exposure to an ultra-low dose of Roundup herbicide" *Scientific Reports* **7** Article number: 39328 (2017)

Miller, G.H., et al Abrupt onset of the Little Ice Age triggered by volcanism and sustained by sea-ice/ocean feedbacks. *Geophysical Research Letters*, 2012

Milne, A. "Earth`s changing climate; the cosmic connection" Prism Press 1989

Moffa-Sánchez,P., Andreas Born, Ian R. Hall, David J. R. Thornalley, Stephen Barker "Solar forcing of North Atlantic surface temperature and salinity over the past millennium". *Nature Geoscience*, 2014

Moon, M. "Japan`s most powerful X-Ray satellite is dead" April 29, 2016 *Space*

Nahle, N., 'Determination of Mean Free Path of Quantum/Waves and Total Emissivity of the Carbon Dioxide Considering the Molecular Cross Section' (2011), *Biology Cabinet*, (Peer Reviewed by the Faculty of Physics of the University of Nuevo Leon, Mexico).

NASA briefing "Sep 23 2008" Solar Wind loses power: Hits 50 year low".
NASA Facts Changing Global Cloudiness June 1999

Neslen, A. "Green biomass boilers may waste billions in public money". Jan 14 2015, *the Guardian Newspaper*
New Scientist, "Universe`s highest electric current found" June 15 2011

Overland, J.E. Klaus Dethloff, Jennifer A. Francis, Richard J. Hall, Edward Hanna, Seong-Joong Kim, James A. Screen, Theodore G. Shepherd, Timo Vihma "Nonlinear response of mid-latitude weather to the changing Arctic". *Nature Climate Change*, 2016; 6 (11): 992

Overland J. and Wang, M."Large-scale atmospheric circulation changes associated with the recent loss of Arctic sea ice. *Tellus,* March 2009

Penn, M. and Livingston, W.,(2010) "long term evolution of sunspot magnetic fields" *IAU symposium* no 273

Petit J.R., Jouzel J., Raynaud D., Barkov N.I., Barnola J.M., et al. June 3 1999, Climate and Atmospheric History of the Past 420,000 years from the Vostok Ice Core, Antarctica, *Nature* 399: 429-436.

Petoukhov, V., Ramstorf, S., Petri, S., Shellnhumer, H.J., (Jan 16 2013), "Quasiresonant amplification of planetary waves and recent Northern Hemisphere weather extremes" PNAS.
Pierrehumbert R.T. (2010) "Principles of planetary Climate", Cambridge University Press.

Pierce, Jeffrey R.: Weisenstein, Debra K.: Heckendorn, Patricia: Peter, Thomas: Keith, David W., (Sep 2010) "Efficient formation of stratospheric aerosol for climate engineering by emission of condensable vapor from aircraft", *Geophysical Research Letters*, Volume 37, Issue 18, 18805.

Popova, E., Zharkova, V. and Zharkov, S "Probing latitudinal variations of the solar magnetic field in cycles 21–23 by Parker's Two-Layer Dynamo Model with meridional circulation "*Ann Geo*. Nov 20 2013

Preston.A., Dutton, W.R. and Harvey B.R., "Detection estimation, and radiological significance of silver 110m in oysters in the Irish Sea and the Blackwater Estuary" *Nature* pp 689-690

Rahmstorf,S. (2003) "The concept of the thermohaline circulation" *Nature* **421** (6924): 699.)

Rahmstorf,S. "Ice Cold in Paris" *New Scientist* Feb 8 1997 p, 26

Rees, J. "Green taxes stoke up steel plant jobs crisis',Northern Powerhouse is dealt a blow as Ministers are blamed for 3,000 redundancies" *Financial Mail on Sunday* Sep 19 2015

G.C.Reid, J.R.Mcafee,P.J.Crutzen, "Effects of intense ionisation events" *Nature* pp 489-492 October 12 1978

Roderick,M. I., Farquhar,G. D., (The Cause of Decreased Pan Evaporation over the Past 50 Years *Science* 15 Nov. 2002: Vol. 298, Issue 5597, pp. 1410-1411

Royal Astronomical Society "Irregular heartbeat of the Sun driven by double dynamo" July 9 2015.

Sagan, C. and Mullen, G., "Earth and Mars, Evolution of Atmospheres and Surface temperatures" *Science* July 7 1972

Sankararamen, S. et al, "The genomic landscape of Neanderthal ancestry in modern day humans" *Nature* vol; 507 p 354-356, 20 March 2014

Schnepp, E. and Hradetzky, H., 1994, "Combined paleointensity and 40Ar/39Ar age spectrum data from volcanic rocks of the West Eifel (Germany): Evidence for an early Brunhes geomagnetic excursion," *Journal of Geophysics Res.*, Vol. 99, p. 9061-9076, 10 May 1994.

Science Daily (2010) Link between solar activity and the UK's cold winters April 19, 201**0 S*ource: Institute of physics***

Samsel, A, and Seneff, S. "Glyphosate, pathways to modern diseases II: Celiac sprue and gluten intolerance" *Interdisciplinary Toxicology* 2013 Dec; 6(4): 159–184.

Service R.F., *Science online* Jan. 11, 2017 **"**Here's how to improve controversial carbon accounting tool says U.S. science academy"

Science Daily "Ozone at lower latitudes is not recovering, despite Antarctic ozone hole healing." 6 February 2018.

Seager, R., D. S. Battisti, J. Yin, N. Gordon, N. H. Naik, A. C. Clement and M. A. Cane, (2002): "Is the Gulf Stream responsible for Europe's mild winters?" *Quarterly Journal of the Royal Meteorological Society,* 128(586): 2563-2586.

Shaviv and Veiser (2003) "Celestial driver of Phanerozoic Climate" *GSA Today* Vol. 13 no 7 July 2013

Speight, M. R. Henderson, P.A. "Marine Ecology: Concepts and Applications", Wiley- Blackwell 2010.

J.A. Screen, and I. Simmonds, "Amplified mid-latitude planetary waves favour particular regional weather extremes", *Nature Climate Change,* vol. 4, pp. 704-709, 2014.

Smith, C.R. **The** Native People of North America *Arctic Culture Area*

9 March 2000

Stanhill, G.; Moreshet,S. (2004-11-06). "Global radiation climate changes in Israel". *Climatic Change* 22(2): 121-138

Svensmark, H. and Calder, N. (2007) *The chilling stars, a new theory of climate change,* Amazon

Thompson, S.L.: Schneider Stephen H.: Neon Winter, Summer 1986, *Foreign Affairs*, Vol.64, No. 5, pp 981-1005

Tickell, O. (2008)." Kyoto 2: How to Manage the Global Greenhouse".(Zed Books Ltd)

Travis, David J., Carleton, Andrew M. & Lauritsen, Ryan G (2002). "Contrails reduce daily temperature range" *Nature*. **418** (6898): 601.)

Turco, R.P., Toon O. B. , Ackerman T.P., Pollack J.B. , and Carl Sagan "Global Atmospheric Consequences of Nuclear War" *Science* 222 pp 1283-1292 , 1983 .TTAPS

Vihma, T. "Effects of Arctic Sea Ice Decline on Weather and Climate": Finnish Meteorological Institute, *Surveys in Geophysics*, 30 January 2014.

Wigley, T.M.I., "A combined mitigation/geoengineering approach to climate stabilisation" *Science* Sep 14 2006

Wilson Clark, *Energy for Survival: The Alternative to Extinction (Garden City, NY: Anchor Books, 1974), p. 117).*

Woillard, G. M., 1979 "Abrupt end of the last interglacial in northeast France" *Nature* vol. 281 p 558-562 18 October 1979

Woods, T. "Press Release Shrinking atmosphere linked to low levels of solar radiation," LASP Lab. For Atmospheric and Space physics, Aug 26 2010

Wright, H. E. Jr., "Late Quaternary Vegetational History of North America," in Late Cenozoic Glacial Ages, Karl Turekian editor Yale publishing, p. 425-464, 1971

Willerslev et al Genomic structure in Europeans dating back at least 36,200 years *Science* Nov 6 2014

Yang Q., Dixon T. H., Meyers P. G., Bonin J., Chambers D., van den Broeke M. R., Ribergaard M. H., Mortensen "Recent increases in Arctic freshwater flux affects Labrador Sea convection and Atlantic overturning circulation" *Nature Communications* **7**, Article number: 10525 Jan 22 (2016)

Yin,S. "The Brilliance of a Stradivari Violin Might Rest Within Its Wood", *NY Times.* December 20, 2016.

Zharkova,V.V. *et al.* "Heartbeat of the Sun from Principal Component Analysis and prediction of solar activity on a millennium timescale", *Sci. Rep.* **5**, 15689; doi: 10.1038/srep15689 (2015).

Made in the USA
San Bernardino, CA
14 May 2018